D0021530

LIGHT IN
THE
DARKNESS

LIGHT IN THE DARKNESS

Black Holes,
the Universe, and Us

HEINO FALCKE

WITH JÖRG RÖMER

TRANSLATED FROM GERMAN BY
MARSHALL YARBROUGH

HarperOne
An Imprint of HarperCollinsPublishers

Online material for the book and updates by the author can be found at:
http://light-in-the-darkness.org
Instagram & Twitter: @hfalcke
https://heinofalcke.org

HarperCollins books may be purchased for educational, business, or sales
promotional use. For information, please email the Special Markets Department
at SPsales@harpercollins.com.

Originally published as *Licht im Dunkeln* in Germany in 2020 by Klett-Cotta.

First HarperOne hardcover published 2021.

FIRST EDITION

Translation by Marshall Yarbrough

Designed by Terry McGrath

LIBRARY OF CONGRESS CATALOGING-IN-PUBLICATION DATA
Names: Falcke, Heino, 1966- author. | Römer, Jörg, 1974- author. |
   Yarbrough, Marshall, translator.
Title: Light in the darkness : black holes, the universe, and us / Heino
   Falcke, with Jörg Römer ; translated from German by Marshall
   Yarbrough.
Other titles: Licht im Dunkeln. English
Description: First edition. | San Francisco : HarperOne, 2021 |
   "Originally published as Licht Im Dunkeln in Germany in 2020 by
   Klett-Cotta." | Includes bibliographical references and index.
Identifiers: LCCN 2020050836 (print) | LCCN 2020050837 (ebook) | ISBN
   9780063020054 (hardcover) | ISBN 9780063020061 (trade paperback) | ISBN
   9780063020078 (ebook)
Subjects: LCSH: Black holes (Astronomy) | Astronomy.
Classification: LCC QB843.B55 F3513 2021 (print) | LCC QB843.B55 (ebook)
   | DDC 523.8/875—dc23
LC record available at https://lccn.loc.gov/2020050836
LC ebook record available at https://lccn.loc.gov/2020050837

21 22 23 24 25 LSC 10 9 8 7 6 5 4 3 2 1

CONTENTS

Foreword

The release in April 2019 of the image of a huge black hole in the center of a distant galaxy rightly attracted a huge amount of attention and interest. This is the story of how that image came to be made. First, the gathering together of a group of scientists—astronomers, and specialists in radio telescopes, radio receivers, and data processing. Then, persuading the guardians of resources (money, radio telescopes, computing facilities) to release enough for this project. And last came making the observations and analyzing the data, producing the image. This account is written by one of the prime movers of the project, one who has had long involvement with it, and who lived it, day in and day out, for some twenty years.

Having worked on some international projects myself, I judge to be pretty accurate the old adage that the complexity of a project goes up as the cube of the number of partner organizations involved. Different bureaucratic methods, different backgrounds, different languages, different outlooks, and different goals all provide multiple pitfalls for the unwary project leader! This project resulted in a paper with 348 authors in eight observatories on four continents making the observations and analyzing the data. The management of a project of this size, with undoubtedly numerous "prima donnas," is an amazing achievement of its own!

Those of us who worked in x-ray astronomy in the 1970s and 1980s quickly had to accept that black holes existed. Okay, those black holes are stellar mass, tiny compared with the ones at the centers of galaxies; but accepting the existence of black holes, accepting that this bit of physics is correct, was the big step. So having lived for some fifty years with these "beasts," I am underimpressed by the excitement around the image of the black hole at the center of M87. But I am impressed that such a large group of scientists and other specialists can be pulled together and kept working together to achieve this image!

Features of black holes (even if not named as such) have long fascinated us. Think of C. S. Lewis's *The Lion, the Witch and the Wardrobe*. That wardrobe is what we would now call a space-time bridge leading the children to a different world in a different season and at a different time of day. Space-time bridges could be formed if a black hole (which swallows things) could be connected to a white hole (which emits things). The connection was named "a wormhole" by John Wheeler. Alan Garner's *Boneland*, set near Jodrell Bank, also suggests some amazing distortions of space and time, although it does not explicitly invoke black holes, and there have been many books explaining the properties of black holes.

Being an astrophysicist, one cannot totally avoid the big questions like: how (or why) was the universe created, and what is there after the death of the universe, and are there other universes? The yawning maw of a black hole reminds us that the universe is not cozy, and there is an embedded existential challenge. However, deadlines and other mundane things call, there is plenty to be getting on with, and so these questions may not occupy center stage in our minds for long.

The universe appears to defy definitive description and full comprehension—questions like where did it come from, or why did it start, do not seem to have a scientific answer. Some of us believe in a God or even a creator God, some don't. Some of us are Christian, some other religions, some none. However, it seems to me that ultimately we come to a point in our understanding/belief system/theology where we have to say we don't know, or we don't understand. We go on living and working, because we have to, and make the best of it. Some of us are better at living with uncertainty and incompleteness, and such untidiness than others.

Falcke sets out his understanding in the final section of this book, and I admire him for making the effort, and for doing so. But arguably this may be more for him than for us! Belief systems are to a degree "tunable"; we can (and do) tune them to our personalities and to our particular needs.

This very readable book is written in a lyrical style—clearly the author is in love with a wondrous universe.

Jocelyn Bell Burnell

Prologue

And We Really Can See Them

The lights go out in the large press room at the European Commission's headquarters in Brussels. The moment that we've waited so long for, that we've all worked many years and to the point of exhaustion to bring about, is finally here. It is Wednesday, April 10, 2019, 20 seconds past 3:06 p.m. Forty more seconds to go, and then, for the first time, people all over the world will marvel at the image of a giant black hole. It is located at a distance of 55 million light-years from Earth, at the center of the Messier 87 Galaxy—M87 for short. For a long time it seemed that the deep darkness of black holes would remain completely and forever hidden from our eyes, but today that darkness will step out into the bright light of day for the first time.

The press conference has begun, but we still don't have the slightest sense of all that it is to lead to. Humanity's thousand-year journey of discovery, traveling to the very limits of our knowledge; revolutionary theories about space and time; the most modern technology; the work of a new generation of radio astronomers; and my entire life as a scientist—today they will all be brought together in this single image of a black hole. Astronomers, scientists, journalists, and politicians watch transfixed, waiting to see what we here in

Brussels and in other world capitals are about to reveal. Only later will I find out that millions of people around the world are glued to their screens, and that in just a few hours close to four billion people will have seen our image.

In the front row of the hall sit distinguished colleagues and young scientists, many of them students of mine. For years we've worked together in an intense collaboration. Each of them has pushed their limits, going far beyond what they or I could have imagined. Many of them traveled to the most remote regions of the Earth, sometimes risking their lives—all for this one goal. And today the successful result, the culmination of their work, is the center of the world's attention, while they sit in the dark. I would like to thank them all right now, because each and every one of them has helped to make this breakthrough possible.

But the clock is ticking. I feel like I'm in a tunnel, every impression flies past me like the wind past a race car driver. I don't notice the phone in the third row whose camera is pointed at me. The clip turns up later as a "trending topic" on a popular website for kids—between vulgar jokes about the president of the United States' rear end and a famous rapper's latest single. The journalists are tense and alert, and I start to feel tense myself: there is expectation in every eye. My pulse is racing. Everyone is staring at me.

Carlos Moedas, the European Union science commissioner, spoke just before me. "Don't speak for too long," we'd told him. Moedas stoked the audience's curiosity with his remarks, but he finished too early. I have to improvise to fill the time, while trying to hide how nervous I am.

This very first image is to be unveiled simultaneously all over the world. At exactly 3:07 p.m., Central European Time, the image will appear on the giant screen here in the hall. At the same time, my colleagues in Washington, Tokyo, Santiago de Chile, Shanghai, and

Taipei will reveal this image of a black hole, offer comments, and answer journalists' questions. Computer servers on every continent are programmed to send academic papers and press releases out to all corners of the world. Time passes inexorably. We've coordinated and planned everything in advance, with the utmost precision—the slightest deviation would throw everything out of sync, no different than how it was during our campaigns to gather observational data. And now I start stumbling right out of the gate.

I begin with a few words of introduction, while a film behind me zooms ever more quickly, ever deeper into the heart of a giant galaxy. I start with a dumb slip of the tongue. I've gotten light-years mixed up with kilometers—no small matter for an astronomer, but there's no time to fall to pieces now; I have to continue.

The display ticks over—it is exactly 3:07 p.m. From the depths of the infinite darkness of outer space, from the center of the galaxy Messier 87, there appears a glowing red ring. Its contours are faintly visible, they linger, slightly blurred, on the screen; the ring glows. Everyone watching is pulled under its spell, is given some sense that this image, which was considered impossible to capture, has finally found its way to us here on Earth by way of radio waves that have traveled a distance of 500 quintillion kilometers.

Supermassive black holes are outer space graveyards. They are made of fading, burned-out, and dead stars. But space also feeds them gigantic gas nebulae, planets, and still more stars. By virtue of their sheer mass they warp the empty space around them to an extreme extent, and seem to be able to halt even the flow of time. Whatever comes too close to a black hole never breaks loose of its grip—not even rays of light can escape them.

But how can we possibly see black holes, if no light can reach us from inside? How do we know that this black hole has compressed the weight of 6.5 billion suns into a single point on its way to becoming supermassive? After all, what the glowing ring encircles is the profoundly dark blackness of its center, which no light and no word can escape.

"This is the first ever image of a black hole," I say, once it has finally appeared on the screen in all its glory.[1] Spontaneous applause fills the room. All the accumulated strain of the past few years falls from my shoulders. I feel free—the secret is finally out. A mythical creature of cosmic proportions has finally taken on a form and color that everyone can see.[2]

The next day the newspapers will say that we've written scientific history. That we've managed to give humanity a collective moment of joy and wonder. They really do exist after all, these supermassive black holes! They aren't just fantasies dreamed up by crazy science-fiction authors.

The image was only possible because people all over the world, despite all our troubles and all our differences, devoted years to the pursuit of a common goal. We all wanted to track down the black hole, one of physics' biggest mysteries. This image led us to the very limit of our knowledge. As crazy as it sounds, our ability to measure and to study ends at the edge of the black hole, and it is a big question whether we'll ever be able to go beyond this boundary.

This new chapter in physics and astronomy began with generations of scientists before us. Twenty years ago the idea of capturing an image of a black hole was still considered a far-fetched dream. Back then, as a young scientist on the hunt for black holes, I stumbled into this adventure, and to this day it has kept me fascinated.

I hadn't the faintest idea how exciting it would be, how it would determine and change the course of my life. It became an expedition to the ends of space and time, a journey into the hearts of millions of people—even if I myself was the last person to understand this. With the world's help we managed to capture this image. Now we were sharing it with the world, and the world embraced it, more wholeheartedly than I would ever have thought possible.

For me, it all began almost fifty years ago. Ever since I first looked into the night sky as a young boy, I dreamed of the heavens as only a child can. Astronomy is one of the most ancient and most fascinating branches of science, and it is still giving us dramatic new insights even today. From the beginnings of astronomy to the present day, astronomers, driven by curiosity and necessity, have continued to fundamentally change our view of the world. Today we explore the universe with our minds, with mathematics and physics, and with ever more sophisticated telescopes. Armed with the most modern technology, we set out on expeditions to the ends of the earth and even into space in order to study the unknown. In the unfathomable reaches of outer space, in the infinite universe, and in the divine cosmos, knowledge and myth, faith and superstition have always been so tightly interwoven that today not a single person can look into the night sky without asking themselves: What still awaits us in this dark expanse?

About This Book

This book is an invitation to take a personal journey with me through this—through our—universe. We begin in part I on Earth, fly past the moon and sun that mark our seasons, days, and years, pass by the planets, and learn from the history of astronomy, which continues to shape our perception of the world today. The second part of the book is a journey through the development of modern astronomy. Space and time become relative. Stars are born, die, and sometimes become black holes. Finally, we leave our Milky Way and keep going until we see before us an unimaginably large universe, teeming with galaxies and monstrous black holes. Galaxies tell us of the beginning of space and time, the Big Bang. Black holes represent the end of time.

The first image ever taken of a black hole was a major scientific effort that involved hundreds of scientists working together for many years. The idea for the image, which grew from a tiny mustard seed into a large-scale experiment; the exciting expeditions to radio telescopes all around the world; and the nerve-wracking time spent working and waiting until finally the image saw the light of day—my own experiences from this adventure I relate in part III.

Finally, in part IV, we dare to pose a few of the last big questions still facing scientists today. Are black holes the end? What happened before the beginning of space and time, and what happens at the end? And what does this knowledge mean for us tiny humans here on this unremarkable yet miraculous planet Earth? Does the triumph of natural science mean that we will soon be able to know, measure, and predict everything? Is there still room for uncertainty, for hope, for doubt, and for a god?

PART I

Journey Through Space and Time

*A brief survey of our solar system
and the early history of astronomy*

HUMANKIND, THE EARTH, AND THE MOON

THE COUNTDOWN

Let us set out together on an exciting journey through space and time. We'll start on Earth, where a rocket towers over the green landscape, an awe-inspiring sight. Birds flap cluelessly around this masterwork of engineering. It's the pregnant silence before the storm; the darkness of the just-budding dawn lingers over the launch site. Nature as yet suspects nothing of the hellish inferno about to be unleashed just a few seconds from now.

Tired but excited, the staff and observers gather on the observation platform. From here every object, every person, indeed the whole scene looks cute, as if it were playing out in a dollhouse. One of the observers takes out his phone and starts streaming the event on a website covered in Chinese characters and flashing logos. It's this stream that I watch online, grateful and full of hope, while I sit on the other side of the Earth in a comfortable bed and breakfast in the green Irish countryside. I watch transfixed as the events unfold.

Suddenly, from somewhere off-screen, a voice starts blaring. It's choppy and unintelligible, metallic-sounding, enough to make your skin crawl. Monotonously it begins intoning a countdown, and although it's in a language I don't understand, I count along with it. With a roaring crash a reddish-yellow light at the base of the rocket illuminates the darkness. The ignition of the propulsion unit makes a deafening noise even here in idyllic Ireland—even though the sound is only coming from my laptop. The ground shakes, the rocket's mounts have fallen off, it breaks free and rises majestically, leaving a glaring trail of heat behind it like a reverse comet before it disappears from view and shoots out into space.

I feel like I'm back at the launch of the space shuttle *Discovery*, which I was able to watch with my tired but excited family from Cape Canaveral in the early morning of February 11, 1997. Still today I can picture the proud look on my four-year-old daughter's face when she saw the towering rocket from afar the day before. In the gleam in her eyes I recognized the gleam in my own.

Twenty-one years later, on May 20, 2018, I'm only watching a pixelated, jerky livestream from China. Nevertheless I know exactly what it must feel like to be there now, and this launch is particularly special. The rocket is carrying a piece of me on board: an experiment by my team in Nijmegen in the Netherlands. I feel just like a kid again. The rocket has a special destination—the far side of the moon.

In my mind I'm flying with it, to the moon and far beyond, just as I've done many times before. I fly where my longing has always pulled me: into outer space.

IN SPACE

Celestial calm. The first thing you notice when you arrive in outer space is the utter stillness. The engines have shut down; outside all sound dies away. The Hubble Space Telescope floats 550 kilometers above the surface of the Earth—almost 70 times the height of Mount Everest. The telescope glides through an atmosphere that is about 5 million times thinner than the atmosphere on the Earth's surface.[1] Sound waves, actual vibrations in the air, are no longer audible to human ears: not a rustle, not a word; not even the most violent explosion on Earth could be heard up here.

As an astronomer, I use the space telescopes that orbit the Earth, listen to the stories told by the astronauts that have been up there, and look at the images they've brought back. In my head I'm floating quietly in space, seemingly weightless, but actually I'm speeding around the Earth at a breakneck 27,000 kilometers per hour. My strong centrifugal force could potentially fling me out of orbit, but the powerful pull of the Earth's gravity evens this force out and keeps me on course. This is the secret behind all orbital motion around a celestial object. Weightlessness doesn't mean you're free from gravity's pull. In orbit, gravity still has us in its grip, but we feel weightless because the centrifugal force and the force of gravity are perfectly balanced. Actually we're in free fall, but we keep missing the Earth again and again because we're circling it on a wide trajectory, so neat it could have been drawn with a giant compass. If we were to slow down, our trajectory would get ever smaller and steeper, until eventually our free fall would end abruptly in an impact crater on Earth. But of course no one wants that!

The scant atmospheric friction with which our spaceship has to contend is so minimal that we can orbit the Earth for years nearly unimpeded,[2] without firing our rockets even once.

So long as we're orbiting in space, we can enjoy the one-of-a-kind view of Earth from up here. Godlike, we look upon this blue pearl, set against the black velvet backdrop of the universe. Continents, clouds, and oceans unleash a rich, wild play of colors. At night, flashes of lightning, glowing cities, and the shimmering auroras light up the global stage—a spectacular sight. Borders disappear; with our all-encompassing view we see the Earth as the shared home of all humankind. The line that divides us from the cold of space is clear and sharp. Only now, from up here, do we understand how thin the layer of air is that protects us from hostile space and makes life possible. The weather and climate play out in just a small strip above the Earth. How fragile, how vulnerable this proud planet seems all of a sudden! We owe such fascinating sights—and insights—in space to modern technology. But through its reckless use on Earth we're also destroying the very basis for our lives on this unique blue planet.

Every time I see these beautiful images of Earth, I also sense the loneliness and emptiness, the pain and suffering that are felt all over the world. "He spreads out the northern skies over empty space; he suspends the earth over nothing," so cried sorrow-stricken Job millennia ago.[3] The nothingness of heaven, spread out like a black canvas, and in the middle—our planet Earth! The biblical writer was not granted this view from above, and yet in his visions he already perceived the Earth as a whole. Humanity's old visions are today filled with new images, provided to us by modern technology. A swarm of satellites with cameras and sensors trained permanently on our planet captures clouds, continents, and oceans in breathtaking detail.

Job, who sees the Earth hanging upon nothing, submits his grievance to God. What Job experiences is something profoundly human: pointless suffering. Still today this planet is a complex mix of suffering and beauty. An individual human cannot be seen from space. Suffering can only be grasped from up close; from afar, everything on Earth looks sublime and extraordinary. Even hurricanes, floods, and forest fires take on a morbid fascination from up above. In space one is far removed from the suffering of individuals, which plays out by the billions down below. From space our earthly problems are incomprehensible. Doesn't this "omniscient view" often look past humans themselves?

It is more than astounding how this sober and technical research can cause such a lasting change even in hardened space travelers. After the cosmonaut Yuri Gagarin became the first in 1961, more than 550 people have been in space. Almost all have reported that their amazement at the sublime fragility of the Earth left a deep impression on them, left them profoundly changed as individuals. The experience of gazing upon the entire globe seems akin to an ecstatic state. The author Frank White called this phenomenon, which he studied and described in psychological detail, the "overview effect." What does the sight of the planet trigger within us? How does it change us? How can we make use of this effect? Doctors have been researching the "overview effect" ever since it was first described. The Earth is unique; there is nothing in space comparable to it, so far as we know. Astronauts have the same impression. Floating above the Earth like an angel and seeing everything from above doesn't leave us humans cold. Let us therefore be inspired by these new images from and of space, without overlooking the human individual.

TIME IS RELATIVE

As soon as we've reached orbit, our perspective of space and time changes. We don't just get a different view of our home planet, Earth; the way we perceive days, months, and years changes as well. "A thousand years in your sight are like a day that has just gone by,"[4] as a verse in a famous old psalm would have it. Time is relative. People have suspected this since the dawn of time itself, but nowhere do we experience it more drastically than in outer space.

When I wrote my first observation program for the Hubble Space Telescope, I had to divide the command sequences into 95-minute blocks, because that's how long it takes the telescope to orbit the Earth. Every 95 minutes the sun rises and falls. For the telescope a day lasts 95 minutes. The astronauts in the International Space Station also experience sunrises in 95-minute intervals, and I experienced them at my desk as I prepared my observations and floated through the universe in my head.

But the relativity of time means more than just having a different measure for the length of a day. In space, clocks run differently than they do on Earth—even if hardly anyone thinks it possible. At an orbit of 20,000 kilometers above the Earth they run 39 microseconds faster per day. Thus in 70 years our Earth clocks will be a second slower than our space clocks. That doesn't sound like much, but today we have no problem measuring this minimal difference. This seemingly unremarkable discrepancy reveals a key aspect of Albert Einstein's general theory of relativity: time really is relative. This theory doesn't just describe our solar system, but also black holes and the space-time fabric of the entire universe.

The path to this discovery was an extraordinarily long one. It begins, in broad terms, with fundamental discoveries like the structure of our own solar system and the laws that govern it, and extends to our understanding of the structure and laws of the entire cosmos. In narrow terms, this path to discovery begins with understanding light's paradoxical way of behaving as both a wave and a particle, and is naturally bound up with Einstein's famous theory of relativity.

The key to all of this is a precise understanding of the remarkable qualities of light. What is especially astounding is that light doesn't just make it possible for us to see, thereby enabling our discovery of the Earth, moon, and stars. In fact, light, time, space, and gravity are all closely interconnected.

Let's take a moment to look back at the history of modern physics. For Isaac Newton, the author of the theory of gravitation, light consisted only of small corpuscles, that is, the tiniest of particles. Later, in the nineteenth century, the Scottish physicist James Clerk Maxwell, using as his basis the brilliant pioneering work of Michael Faraday, developed the theory that light and all other forms of radiation were electromagnetic waves. The radio signals necessary for Wi-Fi, cell phones, or car radios; the thermal radiation picked up by night-vision goggles; the x-rays we use to make bones visible beneath the skin; or indeed the visible light that our eyes perceive— according to Maxwell's theory, these are all oscillations of electric and magnetic fields. They differ from one another only in terms of their frequency and the means by which they are produced and measured. But at their core, these oscillations all represent the same phenomenon—light: radio light, infrared light, x-ray light, and visible light.

In the frequency range used for cell phones, the waves oscillate a billion times per second, and their wavelength stretches over 20 centimeters. Visible light waves oscillate sextillions of times per second and are a hundred times smaller than the diameter of a hair. Because light waves of a certain color and frequency always oscillate at the same rate, light is also the perfect tempo-setter for a clock and the standard measure when it comes to keeping time. The most precise optical clocks today are calibrated to be accurate to within less than 10^{-19} seconds.[5] Over the present life span of the universe, about 14 billion years, such a clock would only be about half a second off! That's a degree of precision that earlier generations wouldn't even have dreamed of.

But what exactly is it that's doing the oscillating? For a long time it was believed that all of outer space was filled with so-called ether. What people had in mind wasn't the chemical solvent, but rather a hypothetical medium in which electromagnetic waves, or light and radio waves, moved and fanned out like sound waves in the air.

One of the aspects of the Maxwell equations that was most surprising and baffling to physicists, and still is even today, was the notion that light of every color traveling in empty space was always supposed to move at the same constant speed, no matter how fast an observer might be traveling. An x-ray was just as fast as a radio wave or a laser beam, and in the Maxwell equations the speed of light was not dependent on the speed of the receiver or sender. We have known that light isn't infinitely fast since the end of the seventeenth century at the latest, when Ole Rømer and Christiaan Huygens measured the movements of Jupiter's moons and used them as clocks.[6] But wouldn't the speed of light have to change if one were flying at high speed through the mysterious ether, or standing still relative to it?

Let's say I'm in the ocean on a surfboard. There's a stiff wind blowing toward land and I'm paddling out perpendicular to the line of surf. The waves are coming toward me at high speed—indeed, just about as quickly as they're crashing on the shore. But if I change directions and surf quickly with the wind and waves, I'm just as fast as the waves under my surfboard. Relative to my surfboard, the speed of the waves is small; relative to the shore, however, the speed of the waves is very high.

The same thing holds true for sound waves. If I'm riding my bike with a tailwind, the sound of a car honking its horn behind me reaches me somewhat faster than it would without wind, and I hear the warning a bit sooner. If I'm pedaling against the wind, the honk from behind reaches me somewhat later; the sound has to travel against the wind. If I were able to pedal at supersonic speeds relative to the wind, then I wouldn't ever hear the honk. If I pedal even faster and outpace my own sound waves, then I break the sound barrier and cause a loud noise, because many of the sounds I make reach the person hearing them at the same time. Unlike jet pilots, however, no cyclist has yet managed to produce a sonic boom.

Radio waves must behave in the same manner, or so people thought more than a hundred years ago. The ether—like air in our atmosphere—fills the emptiness of outer space, and the Earth is like my bicycle or my surfboard, plowing through the ether at 100,000 kilometers per hour on its path around the sun. If you measure the speed of light in the direction of the Earth's movement around the sun, then this "light speed" must actually be a completely different quantity compared to the quantity measured at a right angle or in the precise opposite direction—in other words, it must depend on whether the Earth is surfing through the ether with a tailwind or a headwind.

It was precisely this effect that the American physicists Albert A. Michelson[7] and Edward W. Morley were out to prove toward the end of the nineteenth century. To do so, they measured the relative speed of light in two pipes that stood perpendicular to one another. The experiment was a spectacular failure. They could not prove any significant difference in the speed of light. There was thus no clear evidence that the ether existed—it was just an illusion.

Failures can be groundbreaking, and this failure would become one of a few key experiments that would set the history of physics and astronomy on its current path. That's because the completely unexpected collapse of the ether theory caused whole edifices of theory to teeter, making it possible to cast aside old patterns of thinking and start looking out for new ideas. The best to come along were the new ideas of the young Albert Einstein,[8] who was prepared to radically rethink everything and place physics on a new theoretical foundation. While other physicists were still beating their heads against the wall, Einstein was running headlong into a new era in which space and time no longer existed as absolutes. A bold theory emerged—

the theory of relativity—and with it, Einstein essentially tossed out a physical conception of the world that had prevailed for centuries.

A LITTLE BOY DREAMS OF THE MOON

Having orbited the Earth enough times, we can now initiate the next phase in our space capsule's mission protocol and set a course for the moon. The journey to the moon was an ancient dream of humanity. On July 20, 1969, Neil Armstrong set foot on the moon's surface, taking what might be the most famous step a human being ever took, and the dream became reality. A few years later I could still feel the significance of this moment.

It's a hot summer day in 1971, in the idyllic town of Strombach in the Bergisches Land region of North Rhine-Westphalia. Soft green hills and forests line the horizon, a group of children plays happily on the street in a small neighborhood of private homes. Pails and shovels, a tricycle with a push handle, and a couple of balls are all they need to be happy. The grown-ups sit on lawn chairs in the front yard and look on, relaxed.

But there's one boy, small, a bit chubby-cheeked, who isn't playing with the others. Alone in a dark room he stares captivated at the flickering, blurred black-and-white images on a large tube TV set. *Falcon*, the Apollo 15 lunar module, has just landed on the moon and is transmitting its images back to Earth. After the first spectacular and highly successful space missions, the great excitement that surrounded the moon landings in the Falcke family had quickly fizzled out.

Alone, the young boy can't tear himself away from the screen. Almost five years old, he doesn't yet have any idea of the size of space or the distance that NASA's astronauts had to travel to get to the moon. He can't even imagine how much energy this technological master stroke required or how significant this scientific achievement is. And yet, somewhere deep inside him, he senses how fascinating, how huge this bold undertaking is. This little boy takes in every second of this adventure; every second fires his imagination. What else must be possible, now that a person can walk on the moon, jump up and down on its surface, and even drive a vehicle around (which the astronauts on Apollo 15 actually did)? What else was out there for mankind to discover in this infinitely large sky?

The little boy was me, of course. We were staying at my great-aunt Gerda's house for a few days. Back then the astronauts serving under Commander David Scott's command seemed to me like heroes from the comic books. He and crewmember James Irwin set down in the *Falcon* very close to the Montes Apenninus, one of the largest lunar mountain ranges, while a third member, Alfred Worden, orbited the moon in the command module. When Scott set foot on the surface he said something deeply human: "I sort of realize there's a fundamental truth to our nature: Man must explore!" "Yes!" I thought, "that's me." And today that should be said for all humans.

Like so many children, I wanted to be an astronaut. Later I came to understand, somewhat intuitively, that I wasn't really cut out for it. I was quite well-rounded: I was athletic, I was able to work together with others, I was good in theoretical and experimental work, I knew my way around technology, and I was stress-resistant. But my hands start shaking very easily, and in high pressure situations I make too many mistakes. Many years later I had a chance to talk about this with the German astronauts Ulrich Walter and Ernst Messerschmid at a conference on space travel. They knew what they were capable of without being arrogant about it. "We astronauts have to go through an endless selection process—all the parameters have to be right," one of them said to me. Not all the parameters were right in my case. Nevertheless—my dream to get close to the moon never died.

Depending on where the moon is along its elliptical orbit, a space-ship must cover a distance of between 356,000 and 407,000 kilometers in order to reach it. Most cars don't manage to rack up that much mileage before breaking down, but light needs only about 1.3 seconds to cover this distance. Seen from an astronomical perspective, it's somewhat sobering to realize that even the best cars travel scarcely farther than a light-second—an important astronomical measure.

Light speed is the only truly constant measure in the universe. Thus it makes perfect sense to express the size of outer space in units of light. The light-year is in reality a measure of length, not of time, as one might assume based on the word *year*. We get a sense of the giant distances that exist in space when, in talking about the cosmos, we sometimes speak of many billions of light-years. For astronomers the moon is neither our cosmic front yard nor our cosmic back courtyard. At most it serves as the threshold we cross on our journey out into the universe.

Being a good light-second away also means that everything we see of the moon on Earth is always over a second old. When we look out into space, we always see its past. In the moon's case it's only a little more than a second; in the case of the galaxies that we study, we look back millions and billions of years.

Thus light always reaches us with a "delay"—a small lag for light sources here on Earth and an enormously large one for light from the depths of space. As a result, we can never know exactly what's happening somewhere else at exactly this moment—not in the universe and not even here on Earth.

Incidentally, there's a very practical way to measure and experience the delayed arrival of light from the moon. A Dutch colleague of mine celebrated his wedding in the control room of a radio telescope and sent his "I do" to the moon via radio wave. There it reflected off the moon's surface, and after 2.6 seconds it came back to the control room. This happened so quickly that within that short interval of time the bride wasn't able to run away—the marriage was official. It was probably the world's first "moon bounce" wedding.[9]

For somewhat less celebratory purposes, indeed for purely scientific and technological reasons, we also regularly shoot laser beams at the moon today. These reflect against mirrors that were placed there by the Apollo missions—notwithstanding the claims of conspiracy theorists who say that NASA never landed on the moon. The mirrors function now as they did then. From the delay of the light echo, the movement and distance of the moon can be measured in exceedingly precise fashion and we can test predictions made by the theory of general relativity.

We can also see that every year the moon gets four centimeters farther away from us and the Earth rotates a little more slowly. Gravitational forces bind the Earth and moon together, and tidal forces cause each to slow down the other's rotation somewhat. Each month and day lasts a tiny fraction of a second longer each year. In theory we age somewhat slower as a result, but also die a bit earlier—if our age is expressed in months and days, that is. Four and a half billion years ago there were only six hours in a day[10]—a horrifying thought for workaholics like me.

The rotation of the moon is already almost completely halted. During its orbit around the Earth it turns on its axis exactly once, and as a result it always shows us the same side. That's why the same face of the friendly man in the moon is always smiling down on us. Only since the first moon missions have we been able to see the far side of the moon. It's not the dark side of the moon, as it is often poetically described, because the sun shines there for two weeks every month. Still, the far side of the moon has remained a mysterious and nearly unexplored world.

I've never completely given up on my own personal moon dream, and in a certain respect it has actually come true in the form of the LOFAR radio telescope[11] in the Netherlands, which I led for a time. LOFAR stands for "low-frequency array" and is a network of radio antennas that operate in the low-frequency range. They are linked together to form a single observatory—a supercomputer combines the data and creates a virtual telescope. It is meant to make it possible for us to look back almost to the Big Bang and to help us find all the active black holes in the universe.

Today the LOFAR network consists of 30,000 antennas in various locations throughout Europe—LOFAR has become a continental telescope. But the ideal spot to receive cosmic radio waves without interference is the far side of the moon. This is because on Earth, astronomers' biggest problems are the stray radiation caused by terrestrial radio transmitters and the distorting effect that the uppermost layer of the atmosphere, the ionosphere, has on cosmic radio waves. We never see the far side of the moon from Earth; consequently, you can't pick up any stray radiation from Earth there. "The moon might be the best place on Earth to do radio astronomy," I like to say, half-jokingly. But for a long time, the idea of building antennas there seemed to me an impossible dream.

Both in space travel and in science, you need to have a lot of patience. If you do, sometimes something unexpected happens. I experienced something of the sort in October 2015, when, during a state visit, Dutch king Willem-Alexander and Chinese president Xi Jinping made an agreement to work together on space travel. As part of the agreement, the Chinese offered to take a lunar antenna developed by us for the LOFAR program with them into space. It was the first Dutch instrument included in a Chinese moon mission. In May 2018, a rocket built by the Chinese space agency CNSA launched from the Xichang cosmodrome with this antenna on board. It was this rocket whose launch I followed via livestream while vacationing in

Ireland—at the same time that the very first image of a black hole was coming together. At the time all my energy was focused solely on this image; it was the most strenuous phase of my scientific life. Because of this, I reluctantly had to leave my childhood dream in the hands of my colleagues.

Our LOFAR observation station is mounted on the Chinese communications satellite Queqiao. The satellite, whose name translates to "bridge of magpies" in English, is parked at a distance of between 40,000 and 80,000 kilometers behind the moon. Queqiao's primary function is to relay radio signals back to Earth from the far side of the moon. But in fall 2019 we extended our antenna, and we've been listening to cosmic signals ever since. Most recently we've been searching for an extremely weak radio static that according to present-day theories must have emerged at some point in the dark ages of the universe—billions of years ago, before the birth of the first stars. It contains a radio echo of the Big Bang—the beginning of space and time. We will probably need many years to complete the extremely difficult data analysis, and it's quite possible that only future missions will be able to find something.

But even before it got to the moon, the satellite presented me with a special sight. The small onboard camera managed to take a unique snapshot as Queqiao was on its way to its position in orbit. The photo shows the moon and, behind it, almost the same size, the Earth. In a corner of the photo is our still-retracted antenna.

When I saw this, I felt like that little boy in front of the old black-and-white TV set again: looming there before me was the mysterious far side of the moon, and behind it I saw, tiny and blurred, our own blue planet, on which I was now sitting. I hadn't ever made it to the moon myself, but in this moment I felt "at home" there. Ever since then, every time I look up at the moon, I like to think that there's now a little part of me up there.

THE SOLAR SYSTEM AND OUR EVOLVING MODEL OF THE UNIVERSE

THE SUN: OUR CLOSEST STAR

Once we've left the moon, our next stop is the sun. Traveling from Earth we must cover a distance of 150 million kilometers. Light can manage that in 8 minutes—thus we are 8 light-minutes from the sun. Whoever looks at the sun sees 8 minutes into the past.

The sun is the star that gives us life in almost every respect. Like no other celestial object—except for the Earth—it makes human life possible. It impacts the weather, has a lasting influence on human culture, and orders our daily activities through the rhythm of day and night. We only begin to grasp the importance of the sun when we have to do without it. No wonder, then, that a solar eclipse would cause major alarm among people and societies in prehistoric and ancient times, and that it continues to do so to a small extent today.

It's the summer of 1999 and I'm standing before the principal of our local elementary school, almost begging her to let me take a trip with my daughter. It's the morning of August 11, when a total solar eclipse will darken parts of Germany and France. For days the media has been advertising the event. Special protective sunglasses are sold out; all of Germany is waiting for the cosmic occultation. For my daughter and me it is a once-in-a-lifetime opportunity: by the time the next comparable solar eclipse near us comes around in 2081, I'll no longer be alive.

But the strict rules of compulsory education in Germany don't take such sentimental details into account—our educational laws allow for school to be canceled on days when there's a heat advisory, but not when there's a solar eclipse. The sympathetic principal scratches her ear and informs me that according to school rules she can't let the children out of school even for a once-in-a-century cosmic event—not even the children of astronomers. "However," she adds meaningfully, "the attendance requirement does not apply if you have to change residences temporarily for professional reasons. Then you could take Jana with you." I thank her for the information and change residences for a day—at least on paper.

Excited and filled with curiosity I hop in the car with my six-year-old daughter. Sometimes to be a scientist means that in your search for the mysteries of the universe you want to travel to the ends of the earth to satisfy your curiosity. Now we're starting on our own little expedition.

The umbra of the eclipse will only be visible around midday in a narrow strip covering a few regions in the southwest. That's where I want to be, because only here will it be possible to experience the most fascinating aspect of a total solar eclipse: the ominous gloom that takes over when the world is suddenly plunged into darkness in the middle of the day. Anyone who has experienced this moment will never forget the sense of how important sunlight is for our lives and for life in general. There's only one problem, a problem astronomers are all too familiar with: the weather isn't cooperating. All of Germany is covered in clouds.

We drive west from my hometown of Frechen, just outside Cologne—always on the lookout for the right spot. Desperately we flit about, chasing the sunlight that peeks through the clouds here and there. Finally we end up in France, in a field near the town of Metz. There are now just a few minutes left until the eclipse begins, and right at this moment the sky opens up and the light of the sun shines through. Sometimes in life you just have to get lucky, even as a little scientist. Slowly, majestically, the disk of the moon slides in front of the sun, until finally it darkens it completely. We're at exactly the right place at exactly the right time. It is extraordinary and wonderful. In the truest sense of the word, we are given a rare moment of collective illumination in the middle of the darkness.

A solar eclipse reveals one of our solar system's most remarkable cosmic coincidences. Only because the much smaller moon is positioned at just the right distance from the Earth does it manage to cover the large disk of the sun completely. If it were a bit closer, it would cover more than just the disk of the sun; if it were farther away, there would still be a bright, blinding outline. As it is, however, the moon covers the burning-hot disk of the sun exactly, and lets us see something very special: the sun's corona. It consists of hot gas, several million degrees Fahrenheit, which is sometimes churned up and flung out by gigantic plasma-spewing solar eruptions.

During a solar eclipse, we can see for the span of a few moments that the sun is no sedate star; rather it bubbles and churns like a magic cauldron in a witch's kitchen. But there's something else, something no less magical, that happens amid the large and small explosions on the surface. There the smallest of ghost particles are formed and shot out into space. These are leftover bits of atoms that are split apart in the heat of the sun and afterward zoom through the solar system at high speed. In its nucleus an atom consists of heavy positively charged protons and almost equally heavy neutral neutrons. These are surrounded by one or several shells of much lighter negatively charged electrons.

These energetic, zooming particles are called, somewhat misleadingly, cosmic rays. When they enter the atmosphere, cosmic rays—or rather let's call them cosmic particles—cause, among other things, the spectacle of the northern lights, shimmering and dancing ethereally in the dark sky over Lapland or Alaska. The flood of particles caused by more intense solar storms is also important to us humans for other reasons, however. These storms can destroy the sensitive electronics in satellites, alter the Earth's magnetic field, and hinder the transmission of radio waves. Particularly severe events can even cause excess voltage in our power grids and knock out the electricity supply for entire cities. Fortunately these storms happen only rarely, and thanks to now-regular outer-space weather reports, precautions can be taken in time.

Only during a solar eclipse can we see with the naked eye where cosmic particles originate. The sight has a very special impact on me. From my research I know that the same particle physics that my daughter and I can see at work on the edge of the sun is also taking place on the edges of black holes, though there it takes place on a much more extreme scale. The interplay of magnetic fields and intense turbulence plays ping-pong with these tiny charged particles, flinging them this way and that and loading them with ever more energy. Electrons that are accelerated in this way and diverted within a magnetic field cause the sun, and also the area immediately surrounding a black hole, to shine with radio-frequency light. Cosmic particles that are produced by exploding stars and in the vicinity of black holes reach even higher levels of energy than those produced by the sun and wander through the turbulent magnetic fields of our Milky Way and of outer space.

Some crash into our atmosphere and can be measured. Large-scale experiments, like the one conducted by the Pierre Auger Observatory in Argentina in which I'm still taking part, measure such particles with detectors that are distributed across thousands of square kilometers.

If we didn't understand the physics of the sun and of cosmic particles, we couldn't understand the physics of black holes, either. How astounding is it that throughout the entire universe everything is bound together by the same processes and plays out according to the same laws? The radiation emitted by black holes, the eruptions of the sun, and the auroras on Earth—they're all part of an infinitely interwoven thread of physics that stretches across the entire cosmos.

I feel like I can see all of this right before my eyes during the solar eclipse on August 11. For my daughter it's a nice childhood excursion, with a mix of adventure and curiosity. Afterward she'll make glasses out of aluminum foil for everyone she knows and invite them to look into the sun. What must the neighbors think?

When I look into the sun with my child, I feel awestruck by the forces of the universe. I am particularly taken with the glowing red luminescence of the darkened sun that shines through the thin haze of clouds. This churning ring has something powerful and almost hypnotic about it. It will later inspire me when selecting the colors for the image to accompany our article predicting the radio image of a black hole.

I have the privilege of knowing which cosmic mechanisms bring about a solar eclipse, but people from the Stone Age to the present day have been frightened by the phenomenon. Especially in earlier eras, people were terrified by such events, which they saw and experienced as a message sent by divine forces. Documents dating back more than four thousand years give an account of one such eclipse. Back then, Chinese court astronomers attempted to predict these phenomena with the help of their observations of the heavens. But it didn't always work. According to one ancient legend, two learned men were even killed by edict of the emperor after failing to predict the exact time of a solar eclipse and being drunk when it occurred.[1] Of course, it's quite possible that this famous episode never happened. Today, at any rate, astronomers can predict solar eclipses with precision and without danger. That's not to say that we're not still wrong often enough when we do research at the limits of what is known. Thankfully though, we no longer have to fear the death penalty!

The sun is a star like any other, except of course that it is *our* star and therefore much, much closer and much, much brighter than any other. Neither our moon nor any of the other planets in the sky would be visible without this hot giant, because they only reflect its sunlight. The sun is so huge that it contains more than 99 percent of the mass in our solar system. Its gravity holds our solar system together, and it is this solar system of ours more than anything else that we have to thank for what we've learned about stars and gravitation.

The sun is a massively large and forbiddingly hot ball of gas in which nuclear fire burns. Hydrogen, of which the sun overwhelmingly consists, serves as the fuel. This light element fuses together in the core of the hot star and becomes helium. The temperature in the core is an unimaginable 27 million degrees Fahrenheit. The sun's surface measures a still-considerable temperature of 10,000 degrees Fahrenheit. The emission of this heat is the ultimate source of all our energy on Earth, and it wouldn't be produced without the gravity and the resultant high pressure in the sun's core. Without sunlight, plants can't grow; they derive their energy from photosynthesis. We, too, rely on the sun for our food, no matter whether we're vegans, vegetarians, or meat-eaters, because animals also live off plants that in turn rely on sunlight.

When we burn wood, we burn energy from the sun. Oil, gas, and coal are the leftover remnants of biological processes dating back to the Earth's beginnings—that is, stored solar energy. In just a short amount of time, however, we are destroying all our reserves and burdening the climate with substances and energy that were deposited over millions of years. You don't have to be a climate scientist to understand that if we keep this up it's not going to end well.

Without the sun we also couldn't produce any electricity. It goes without saying that solar energy would never have been invented, but hydroelectric plants also function only because the sun continually causes water to evaporate and produce rain to fill our lakes and rivers. Even wind turbines operate only because the sun warms up our atmosphere and creates regional differences in temperature, which then stir up the wind. Tidal power stations are alone in drawing their energy from the moon, and nuclear power plants take theirs from elements that were created in space with the birth of black holes and neutron stars. Still, though, it was only thanks to the sun's gravity that these elements ever made it to us in the first place. The ultimate origin of all the energy in the sun, moon, stars, and elements, however, is the Big Bang, the universe's original energy source.

The sun accelerated our path to becoming two-legged creatures capable of thinking abstractly. Its cosmic particles, raining down on the Earth, increase the rate of mutation in the cells of organisms. That these cells were able to develop further, that evolution progressed, that humans evolved from small mammals—we owe all this to the sun. We are, in a certain sense, cosmic mutants. Elevated rates of mutation, however, also bring forth cancer cells, and with them death and decay. Our existence as humans has been hard-fought, earned at the cost of deep suffering. But without these potentially dangerous genetic changes we would still be single-celled organisms.

Compared with other, wilder stars, the sun has a fairly calm temperament, and as stars go it's actually just average, neither particularly large, nor particularly massive, nor even especially active.[2] At 4.6 billion years old, it's also in the prime of life. Considering its total mass, the fusion reactor in the core of the sun is burning at low flame. The amount of energy generated per unit volume is significantly lower than the human metabolism. Our body is a well-honed machine that is permanently running at full capacity. If we all stood really close together, we would be a small star.[3]

But thanks to its size, the sun outshines absolutely everything. The entire world population would have to increase almost a quadrillion times to generate as much energy as the sun produces.

The sun practically burns itself. By fusing hydrogen into helium, matter is converted into energy. As a result our star becomes about four billion kilograms lighter—per second. Considering the large amounts of energy that it emits, it uses up only a minuscule part of its own mass and is thus incredibly efficient. No human-built machine to date can produce so much energy with such little fuel. If our bodies were as efficient and economical as the sun, each person would need less than half a gram of food in his or her entire lifetime. In outer space, stars are only outdone by black holes when it comes to efficiently converting mass into energy.

Nevertheless, there's a bit of sad news in all of this: at some point the sun's tank will be empty. Refueling isn't an option. The fire of the sun will burn out, and with it—if we've made it that far—life on Earth. But we're not quite there yet. Current prognoses give the sun another five to six billion years. That's enough time for us to invest in more solar panels!

GODS IN HEAVEN: THE MYSTERY
OF PLANETARY ORBITS

Once we leave the sun and direct our gaze toward the planets that orbit it, the distances we speak of quickly go from light-minutes to light-hours. Here among the planets lies the key to how we came to understand gravity and develop our modern conception of the world. Spacecrafts built by humans have traveled as far as the planets and a short ways beyond. Everything beyond our solar system we are only able to observe with telescopes.

While Mercury, the closest planet to the sun, is only about 60 million kilometers away from it, Neptune, the planet farthest away, plies its orbit 4.5 billion kilometers away, or at a distance of four light-hours. It takes 165 Earth years for Neptune to make one revolution. For millennia our ancestors have watched the planets and marveled at their regular and yet irregular movements. Fixed stars have a set place in the firmament, while we turn beneath them, but planets seem to wander among the stars. Hence their name: *planet* means "wanderer."

In our sky, all the planets and also the sun and moon move along the same strip—as if there were a planetary racetrack. We call this invisible strip of sky the *ecliptic*, from a Greek word meaning "to disappear or fail to appear; darkness." As the root suggests, the term is related to solar eclipses, which are visible in this area.

The ecliptic exists because all the planets revolve around the sun on the same plane. In so doing they form a virtual disk with astronomical dimensions. The Earth's orbit is itself part of this disk, and because we are located within it, it appears to us as only a narrow strip of the sky—like an old vinyl record that we look at from the side. Planets that are closer to the sun travel faster around the sun than the Earth. They have to do this so that their centrifugal force can counterbalance the much stronger gravity of the sun. The closer we are to the sun, the stronger we feel its gravitational pull. Planets farther out move more slowly than the Earth, because the gravity there is weaker. If they were faster, they would break out of their orbits around the sun.

From our perspective on Earth, the planets take curious paths relative to the fixed stars of the sky. They're like sprinters in a track-and-field stadium, where we are one of them. The sprinters in the outer lanes have to run greater distances and are also considerably slower. The planets Mercury and Venus are the top sprinters in the inside lanes. They are especially quick, and are always near the sun. That's why they can only be seen in the morning and in the evening— Venus is the most frequently visible evening and morning star. The large planets are the slower weekend joggers in the outer lanes; our Earth regularly laps them. From our perspective they seem to move backward, until the Earth has passed them and reached the opposite side of the solar stadium, at which point we seem to be moving toward them. Seen from this opposite side, they suddenly seem to be running in the opposite direction.

It took us humans thousands of years to make this discovery. The courses of the planets visible to the naked eye, Mercury, Venus, Mars, Jupiter, and Saturn, remained a mystery for millennia. No wonder they influenced our religions and our different conceptions of the world.

Long before the reasons for these cosmic phenomena were understood, astronomy served very different purposes. People of almost all religions revered the stars and heavenly bodies. It was quite natural for them to do so, because the stars gave order to daily life and the cycles of the year. The sun dominated the day, and the point of its rise and setting marked the course of the year and the seasons. The phases of the moon gave us the measure of the month, which for unknown reasons roughly corresponds to the female menstrual cycle. Sun and moon seemed to preside over fertility and over human fortune and misfortune. Small wonder, then, that we would praise these divine powers.

THE ORIGINS OF ASTRONOMY

The first archaeological clues relating to the study of the sky are tens of thousands of years old.[4] Observers of the heavens produced calendars once they understood the alternating sequence of daytime, nighttime, and the time of year. First, the cycle of the moon was used to mark time; later this timekeeping method was synchronized with the path of the sun. An early European witness of this is the famous Nebra sky disk. The over-3,700-year-old bronze plate is considered the oldest concrete representation of the heavens.[5]

Humans were able to use these astute insights for agriculture or for navigation at sea, which back then was an extremely risky and dangerous undertaking. Today we have navigation satellites, but ultimately their coordinates still depend on astronomical observations—not of the stars, but of the radio emissions of distant black holes, which we've come to use as cosmic landmarks.[6]

Sometime in the third millennium before Christ, educated priests in what would later be the city of Babylon in Mesopotamia regularly kept track of the positions of the moon and the planets. They used the moon as a calendar for their feast days, but also as a way of determining harvest and tax periods. The administrative month had 30 days and the year 360—the missing days were filled with leap days. Their numeric system was based on the number sixty and not ten, like ours. It's likely because of the Babylonians that we divide the day into twenty-four hours and the circle into 360 degrees.

With the development of cuneiform writing it became possible to compare cosmic information independent of the time of its observation. Next, sometime around the first millennium before Christ, there came an exceedingly well-organized program of observation and dramatic advances in the development of mathematics. Between the Tigris and Euphrates there were whole teams of scholars occupied solely with measuring and calculating the goings-on in the sky. Thousands of cuneiform tablets were filled with astronomical data. Suddenly it was possible to analyze astronomical events occurring over generations, not just within the narrow recollection of single individuals. This was the beginning of the practice of carefully recording, archiving, and analyzing data, and can probably already be described as scientific, even if it served primarily religious purposes.

For the Mesopotamians, the universe was orderly, but also subject to the will of the gods. Their plans could be interpreted from omens, such as, for example, from the aspect of the planets.[7] Once observers of the heavens were able to predict the course of the planets, they tried to use this knowledge to interpret the future. Rulers had horoscopes cast for themselves in order to determine the best time for their undertakings.

I can well imagine that the new arithmetic and the predictability of the planetary movements made an enormous impression on people. These advances probably led to the idea that fate itself could be determined. The Babylonian astrology that emerged from these beginnings would influence many cultures. Even in the Bible, the "three wise men" serve as a literary monument to the astrologers of the Orient.[8] It took millennia to find out that astrology ultimately rests on a false assumption: even if the course of a few celestial bodies can be predicted, the same doesn't apply for human life.

In Egypt, the rhythm of time was determined by the Nile floods, which brought fertile silt from upriver. For the Egyptians, the heavens had a mythological basis. The sun, as an aspect of the god Ra, was daily reborn and rose out of the waters to the east. As humans understood it, Ra gave life and kept everything alive. He crossed the sky, descended in the evening in the west, died, and was reborn the following morning. It was an eternal cycle.

Heaven and Earth met on the horizon: a person living at that time, looking above and around him or her, must have been positively filled with the sense of living on a planet in the very middle of the cosmos. The idea that the Earth was flat was widespread back then and lined up with the anthropocentric sense of life. The Egyptians believed in a cosmos with a world above and a world below. Gods were everywhere, and made sure that the entire edifice of the world remained stable and balanced: the earth god Geb reigned below; above reigned the sky goddess Nut, who was the mother of all stars. Between earth and sky was the realm of the god Shu, the god of air and light. He held up the heavens and made sure that nothing fell down onto the earth.

The ancient Babylonians imagined that the land was a disk floating on the ocean that surrounded the world. The gods lived in the sky and determined the course of the stars. The firmament covered the earth like a bell jar. This model influenced all of antiquity and was aligned with the science of the time.

The Greeks also believed in a world above and an underworld. They intensified the practice of observing the heavens, immersed themselves in mathematics, above all geometry, and combined the Babylonians' observations of the stars with the geometry of the Egyptians. As early as the sixth century BC, Greek thinkers like Pythagoras had come to the realization that the Earth had to be round. Plato (b. 428/427 BC) also mentions the spherical shape of the Earth in his writings.

One of the many achievements boasted by the natural sciences of antiquity that continue to impress us today is Eratosthenes of Cyrene's measurement of the circumference of the Earth sometime around 200 BC. In two distant Egyptian cities, he had the angle of the shadows measured at exactly midday. In one city the sun was right at its zenith, thus casting no shadow; in the second town it did, however, cast a shadow. There the surface of the Earth seemed to tilt 7 degrees further. Because Eratosthenes had carefully measured the distance between the two cities and now knew the angles of the shadows, he could use these measurements to calculate the size of the Earth with relative accuracy—an astounding achievement for the time. Knowledge of the Earth's round shape was preserved in Europe into the Middle Ages and early modern era, and was taught at universities.[9] The idea that the learned contemporaries of Christopher Columbus were supposed to have believed that the Earth was flat is a myth—like so much that people like to say today about the medieval period we dismissively call Dark Ages.[10]

It is true, however, that it would not have been possible back then to convince either rulers or the common people that the Earth wasn't at the center of the cosmos. Since the beginning of human thought, the universe had been the home of the gods and planets. For the Babylonians, even the division of the week into seven days originated with the seven heavenly bodies that we can see with the naked eye: the sun, the moon, and the five closest planets. The Romans renamed the planets after their own gods: Mercury, Venus, Mars, Jupiter, Saturn—all from the Roman pantheon. In European languages, the names of the days of the week are derived from these gods' names—to varying degrees, depending on the language.[11]

The Greek thinkers influenced our conception of the universe for a long time, aided in no small part by the almost overwhelming authority of Aristotle (b. 384 BC), who from antiquity until well into the Christian era was considered *the* philosopher. His influence was so great that any other view seemed preposterous. Aristotle wasn't an astronomer, and his model of the universe was relatively simple. After his death, however, significant astronomers of late antiquity like Hipparchus (b. 190 BC) and Claudius Ptolemaeus (b. 100 BC), better known as Ptolemy, expanded upon it. Still, the Earth continued to form the center of the cosmos. All the planets and stars in the spheres of heaven circled around a single point in the center, the Earth. Ptolemy compiled all the knowledge of the astronomy of antiquity into his *Almagest*, a thirteen-volume comprehensive work that solidified what came to be known as the Ptolemaic system. It's true that individual scholars like Aristarchus of Samos (b. 310 BC) believed in a heliocentric model, in which the sun and not the Earth stood in the center of the universe—but in the end the geocentric model triumphed.

A NEW MODEL

As unimaginable as it is today, this conception of the universe lasted for about 1,500 years. It was just as well regarded by capable astronomers in ancient China, in India,[12] and in the Islamic-Arabic world as it was in the Christian European world—until Nicolaus Copernicus and Johannes Kepler revolutionized it. As theologians versed in mathematics, they eventually reached the point where they would no longer be led astray by the authority of the ancient philosophers.

A few years ago a work trip led me to Beijing for the twenty-eighth general assembly of the International Astronomical Union, the IAU. Thousands of astronomers from all over the world had traveled there to meet, discuss the results of the latest scientific research, and decide on resolutions, such as the naming of celestial objects. At this assembly a local scientific historian gave a lecture on the history of astronomy in China. China's astronomers watched the sky for thousands of years and, even in ancient times, could rely on substantial financial support. This resulted in an impressively large treasure trove of observational data being compiled over a long period of time, data that are still used even today. Until the eleventh and twelfth centuries AD, Chinese astronomy was far more advanced than astronomy in the West. Nevertheless, according to the historian, China at that time did not produce a scientist with the mathematical capabilities of a Copernicus or a Kepler. The Chinese astronomers did too little with their data.

"Why?" asked someone in the audience. "It might have had to do with their conception of the world," the speaker surmised. While in the West many thinkers were beginning to look for scientific explanations for the mysteries of the heavens, in China the focus was primarily on the supernatural. The world was a complex organism, the heavens full of spirits and mythical beings. Everything was too closely interwoven, a much different concept than that of a single, distant, and all-powerful god and creator, which prevailed in the West.[13] For Chinese astronomers, it made no sense to ask what had set the stars in motion. In the West, on the other hand, the ancient polytheistic beliefs had been suppressed to a greater and greater extent by the monotheistic Judeo-Christian worldview, even if superstition and pagan beliefs, along with astrology, never completely went away.

Judaism was also strongly characterized by rational argumentation. The interpretation of the Torah, the Holy Scripture, was conducted by means of intense debate, painstaking demonstration of evidence, and logical chains of argument. Interestingly, in contrast to other religions, the astronomical conception of the world didn't play a particular role in Jewish tradition. True, this tradition did develop against the background of the knowledge of the cosmos available in the Orient at that time, between Babylon, Greece, and

Rome, but in the creation story told in Genesis, the sun, moon, and stars are downgraded, becoming mere "lights." In this grand story, appearing at the very beginning of the Old Testament, our present-day world emerges step-by-step—divided up by the narrative into daylong sections: in the beginning there is light, then the waters and dry land are gathered, and at the end, plants, animals, and man appear. Lights in heaven don't appear at the beginning of creation, but rather—almost coldly diminished—somewhere in the middle. They aren't divine beings; rather they're only there to give time to us, dividing the day from the night. Genesis describes a highly rational world, purged of magic. In the world of the Bible, miracles are very much the exception.

Thus in the Judeo-Christian worldview there is nothing supernatural about nature. It doesn't have a will of its own but was formed by a single God who is the creator and origin of all things, who always was, always is, and always shall be. In this conception we discover an important foundation for the modern natural sciences, namely the ability to rely on a set of principles that underlie nature. Only when you have this assumption does science begin to make sense.

It's true that you read again and again that faith and science are trapped in an eternal struggle, but this is a myth[14] that the age of secularization that began in the nineteenth century is all too happy to propagate. Historians today take a far more nuanced view.[15] The sciences were for a long time a part of theology, not disciplines in their own right. The monasteries of the Middle Ages were strongholds of knowledge and its dissemination; universities were founded with the church's blessing. Many significant scientists received an education in theology and were deeply pious and often employed in the service of the church. Nevertheless, in all branches of science the church claimed final say on matters of interpretation, which in the fifteenth and sixteenth centuries led to more and more conflict.

The Renaissance and Reformation had long since taken hold of people by this point and had fundamentally changed—one could also say revolutionized—their conception of the world and the individual's role in it.

The cosmological revolution began in 1543 with the bold drafting of a new cosmic model—though it wasn't entirely new, of course—by the Prussian-Polish canon Nicolaus Copernicus. In it, the sun moves to the center of the cosmos, and the Earth rotates on its own axis. Like all the other planets it moves in orbit around the sun. Mathematically, this was convincing and forward-looking, but what was baffling was that in this model the universe had to be much larger than previously thought, and the Earth had to turn at great speed. If we were really rotating at such a breakneck speed, wouldn't we have to notice it somehow?

It would take a long time for the new model to become established. Even learned contemporaries of Copernicus, no matter whether they served the church or secular rulers, had good reasons to doubt it. The influential Danish astronomer Tycho Brahe didn't believe in a great mysterious force that caused the world to rotate. But he also knew that the Ptolemaic model of the cosmos couldn't be accurate. And so Brahe, an excellent conductor of observations, left behind crucial data, which the German mathematician, theologian, and astronomer Johannes Kepler later used to derive his famous laws of planetary motion. Using Brahe's observations, Kepler discovered that planets orbit the sun on an elliptical course and not a circular one, and that a planet travels faster the closer it is to the sun. For Kepler, who was intent on finding the beauty and harmony of God in the cosmos, the elegance of his mathematical equations was a satisfying discovery in a theological sense as well—after all, it accorded with the constancy of the creator, who went about his work like an architect drafting a blueprint.

Brahe's foundational measurements were likely the last great achievement in astronomy that were managed without the use of a telescope. The telescope was invented in the early seventeenth century by a clever optician from Middelburg in the Netherlands[16] and initially saw use in ocean-faring until Galileo Galilei thought to point it at the starry night sky over Padua. When, in 1610, he discovered the first moons of Jupiter, the discovery triggered intense debate in Italy and throughout Europe. The Italian Galilei grew more certain of the rightness of the Copernican model, because the fact that the newly discovered moons revolved around Jupiter showed definitively that not all celestial bodies revolved around Earth.

Soon the young scientist began to grow ever more confident. Long patronized by the Catholic Church, his theories were discussed favorably by the Jesuits at first. But in his writings the ambitious Galilei ignored the works of Kepler and continued to believe in the circular orbit of the planets. As a result, strictly speaking, his model didn't accord with the best data available in his time. With his brash manner, the deeply pious Galilei, who in making his claims went so far as to question the authority of the pope, overtaxed the pontiff's initially favorable disposition. Around this time even the writings of Copernicus were placed on the Index of Prohibited Books and subsequently could only be published with a dozen changes. In 1632, there was a showdown before the Inquisition in Rome. Galilei was placed under house arrest for the remainder of his life, but continued to receive financial support from the archbishop of Siena. It then proved difficult to publish his writings in Italy, for which reason they appeared in other parts of Europe.

Galilei was a good communicator and rhetorician who understood how to extend awareness of the results of his research beyond expert circles. At the same time, however, he neglected to give proper credit to the work of other scientists. Today many legends and stories proliferate around Galilei, not all of which stand up to close historical inspection; rather, they contain a lot of modern projections onto both him as a person and the times in which he lived.[17]

After Kepler and Galilei, it took a full two hundred years until the last scientific arguments against this new model were exhausted. The rethinking, however, had long begun.

Looking back from today's perspective, it seems to me that the achievements of Johannes Kepler are the more seminal. Kepler was by nature the opposite of Galilei: an excellent mathematician, slight in stature and prone to illness, he was plagued by self-doubt all his life and suffered a continual series of misfortunes in his personal life. His mother was accused of being a witch by the governor or *Vogt* of Leonberg,[18] and when his wife died he had a hard time with the arduous search for a new partner—he never had much luck with women. But, still today, his three Keplerian laws form the foundation for celestial mechanics. It is from them that the mass of stars and the existence of dark matter is deduced. When I'm explaining black holes in my lectures today I start by mentioning Kepler's law governing the motion of planets around the sun. Matter travels around a black hole in almost exactly the same way—only much faster.

Upon Kepler's foundation, the English theologian and polymath Isaac Newton[19] would manage not only to establish classical mechanics fifty years later, but also to explain the operation of gravity on Earth, in the orbit of the moon, and in the motion of the planets around the sun using his laws of gravitation.

In Newton's model, gravity is a universal, far-reaching force that ensures that bodies of mass exert an attraction regardless of their composition. The force has a weaker and weaker effect the farther apart the bodies of mass are, but it is never completely absent. Newton's force of gravity applies equally and instantly to all bodies in the entire universe—for planets just the same as for falling apples on

Earth; for the ocean's tides in general and the spring tides caused by a full moon. With Newton, almost the entire solar system was explained. But only almost.

VENUS, GODDESS OF LOVE
AND YARDSTICK OF THE UNIVERSE

The size of the universe and the distance between the Earth and the stars remained a fundamental and long-unresolved question in the study of the heavens. If the Earth revolved around the sun, didn't that mean the positions of the stars in the sky would have to shift?

This apparent shift in the location of the stars is called parallax and occurs when a star is observed from two positions located far apart from each other. Anyone can easily perceive the effect for themselves: hold your arm straight out in front of you, stick your thumb up, and look at it first with one eye closed, then with the other. Seeing the thumb from these slightly different perspectives has the effect of making us think the thumb is moving from side to side. The closer we pull our thumb toward us, the greater the apparent movement becomes. When we look at an object far away from us with each eye, we can perceive depth and are thus able to gauge the distance.

What holds true for our eyes at a small scale also holds true for the path of the Earth around the sun. If I measure the location of a star once in the summer and once in the winter, when the Earth is positioned first to the far left and then to the far right of the sun, then the stars, too, depending on how far away they are, should also shift to the right or left. But back then no one saw anything of the sort. Either Kepler and Copernicus's model wasn't accurate, or the stars had to be so far away that the shift was minimal and scarcely perceptible. How far away they were, and thus the size of the visible universe, depended on the exact distance between the Earth and the sun. Determining this became one of the great challenges in astronomy. Tackling it required coordination among astronomers all over the world—though it also meant a global competition.

The object of their desire became Venus, named after the Roman goddess of love and beauty. In reality our neighbor planet is very hot and not very enticing. The pressure of the dense atmosphere of greenhouse gases that surrounds it would crush us. On the surface of Venus the pressure is as great as that at a depth of more than 900 meters underwater on Earth, and it's as hot as an oven to boot.

But Venus rendered an invaluable service to modern astronomy. With the help of this planet, it was possible to measure the exact distance between the sun and Earth—a distance that became known as the astronomical unit (AU)—and along with it the size of the solar system and even the extent of the entire universe. To do this, researchers needed a so-called Venus transit, the short span of time in which Venus passes directly in front of the sun, similar to a solar eclipse, only much smaller and only visible to trained astronomers with a telescope.

While the moon, as a result of its proximity to Earth, sometimes covers almost the entire sun, the far more distant Venus doesn't manage this. Only a smaller, much less noticeable spot can be observed when the planet drifts past the bright golden sun over the course of a few hours. For a long time we humans didn't even register these events.

As early as the seventeenth century, Johannes Kepler predicted transits for Venus and Mercury, the two planets between the Earth and the sun. He didn't live to see his predictions confirmed, however; Kepler died before he was able to observe the next Venus transit in 1631.

The track that Venus's shadow follows across the sun depends on the place of observation on Earth and the distance to the sun. The farther south one travels on Earth, the farther up the shadow moves, because the angle at which one sees it changes. By measuring the time it took Venus to complete its transit from different locations on Earth, it would be possible to calculate the distance between the sun and the Earth using Kepler's laws—a brilliant idea. There was just one catch: Venus transits are astronomical rarities. The main reason for this is that the orbits of Venus and the Earth are tilted slightly away from each other. Even if Venus, seen from the Earth, lies in the same direction as the sun, it can travel above or below the disk of the sun as it moves past. Venus transits occur only four times every 243 years, and they come in pairs—most recently in 2004 and 2012, and before that in 1874 and 1882.

Even if all the conditions were right, the scientists had to make sure not to miss the passage of the planet under any circumstances. Again and again, astronomers from several different countries set out into the world to follow the course of Venus from every possible angle. In a certain respect they were the forerunners for our modern black hole expeditions. Back then such undertakings were anything but easy. Some came close to failure even without leaving home: in England, Jeremiah Horrocks almost missed the Venus transit of December 4, 1639. Initially, he waited next to the telescope that he had trained on the sun. Since Venus wasn't yet visible, Horrocks left his post and went, we think, to attend a church service. When he got back home, it was almost too late. The transit had long begun, and Venus was already in front of the sun. Horrocks could only estimate the full duration of the transit.

Scientists wanted to observe the next Venus transits of 1761 and 1769 far more closely, and so they set off on various international expeditions. But that wasn't so easy either. The most dramatic failure was experienced by Guillaume Le Gentil, who planned to observe the transit from India. When his ship arrived at its destination of Pondicherry in the southeast of the country, the British had just captured the city in the course of a military campaign. The Frenchman Le Gentil couldn't enter, and he was forced to take his measurements on board. But a wooden ship pitching and tossing at sea isn't a good place for precise astronomical measurements; the result was unusable. Le Gentil decided to wait eight years for the next transit, but right when the big moment was approaching, the sky clouded over. Part of astronomy is having the weather on your side at the right moment. Not everyone gets so lucky. When the Frenchman was finally about to sail home after many years abroad, he got sick and almost died of dysentery. Back in France his family had long since assumed he was dead and divided up his property. Even his seat in the French Academy of Sciences had been given to someone else.

Nevertheless, the scientific community was ultimately able to come up with a passable result for the distance between the Earth and the sun. The value of the astronomical unit determined back then differed from today's established value of 149,597,870,700 meters by a margin of only about 1.5 percent.

It wasn't until 1839 that the German astronomer Friedrich Bessel became the first to determine the exact distance to 61 Cygni, a star, with the help of parallax and the astronomical unit. The shift of the star in the sky, which Bessel measured over six months, came out to a minuscule 0.3 arc seconds, the equivalent of a hair's breadth seen at a distance of 50 meters. With simple trigonometry and the astronomical unit he could now calculate the distance from the Earth to the star: it was 100 trillion kilometers—11.4 light-years. Bessel realized with astonishment that the starlight measured by him had taken more than a decade to travel to Earth. With this measurement, the last of the original scientific objections to the heliocentric worldview were finally overcome.

Because almost all the distances in astronomy are deduced via parallax, astronomers came up with a special measure of length in its honor: the *parsec*. The word stands for "parallax second" and equals the distance at which a star would exhibit a parallax of one arc second. This is about 3.26 light-years. The parsec is thus not a measure of time, as some *Star Wars* films might lead you to believe,[20] but rather a measure of length.

The star closest to us is Proxima Centauri. It is 4.2 light-years away—or 1.3 parsecs. That means there is not a single star within a parsec of the sun. Today, with the help of the European space probe *Gaia*, we can measure the parallaxes of almost two billion stars in our Milky Way at distances of up to a few thousand light-years away. With global networks of radio telescopes we can measure the effect for a few stars and gas clouds up to more than 60,000 light-years away, on the other side of the Milky Way.[21]

If satellites today cross the solar system with what seems like a sleepwalker's ease, or astronomers measure the extent of the universe, we still owe these successes to earlier astronomical expeditions from the seventeenth, eighteenth, and nineteenth centuries, which explored our solar system with nothing more than the first telescopes and a few bold ideas. None of these astronomers set off on their own. The heavens belong to all of us, and sometimes you need the entire world to study them. Global collaboration and competition have always been a part of the very nature of astronomy. From the first astrologers of the Orient back in Biblical times, to the study of the solar system and the expeditions to measure the Venus transit, right on up to the attempts to detect gravitational waves or produce radio images of black holes, astronomers have set out into the world and the universe again and again, working both side by side and in competition with one another, with the goal of exploring and measuring the heavens.

PART II

The Mysteries
of the Universe

*A journey through the universe as we
know it today, and through the history of
modern astronomy and radio astronomy:
the revolution sparked by the theory of
relativity, the birth of stars and black holes,
the mystery of quasars, the expanding
universe, and the discovery of the Big Bang*

EINSTEIN'S HAPPIEST THOUGHT

LIGHT AND TIME

The sun is the brightest light in our sky, and the size of the solar system is the fundamental measure for astronomy and our universe. We measure distances in the solar system in increments of light: light-seconds to the moon, light-minutes to the sun, and light-hours to the outer planets. But in our everyday life as well we use light to measure distances for all sorts of things without even knowing it. Until 1966, all units of length were determined by the International Prototype Meter. This was a platinum-iridium bar that was stored in Paris and served as the standard measure. The International Prototype Meter equaled one ten-millionth of the quarter circumference of the Earth as measured along a meridian running through Paris from the north pole to the equator. No wonder the British never really adopted the metric system. Today the meter is based on the speed of light and equals the exact distance that light travels in a vacuum in 1/299,792,458 seconds. Why the unwieldy number? Well, it is the same exact length as the International Prototype Meter in Paris but is no longer bound up with national pride. Whoever uses a meter stick is in reality measuring in increments of light.

Because it consists of electromagnetic oscillations, we also use light for timekeeping. Light has indeed become a fundamental measure of things, and this is true in the most profound sense. Einstein asked himself what it would mean if light always traveled at the same speed no matter how quickly we ourselves were moving. This line of thought would turn all our notions about space and time being immutable and absolute on their head.

But how can light always travel at the same speed? An ant that is crawling inside a fast-moving sports car moves faster than an ant that is simply walking along the asphalt, since the speed of the car is added to the speed of the ant. Shouldn't it be the same with light? No, because light isn't an ant, or a car, or a football, or a rocket. Light is pure energy—and light has no inert mass. Matter can only be accelerated through the application of force and energy. But the lighter it is, the easier it is to accelerate. It's easier to accelerate an ant than it is a car. Light is so "light" that you don't even need to give it a push; light takes off of its own accord. That's why, in empty space, it always travels at the maximum speed, namely, light speed—almost exactly a billion kilometers an hour.

Nothing can move faster than light, because nothing can be less inert. Even changes in gravitation and the gravitational waves that are created as a result can only be produced at light speed. What once began as the speed of light has now actually become a speed of causality. When we speak of "light" here, we're often tacitly including other processes that convey light-like information via massless waves.

But surely something has to change when you move relative to light, right? Yes, said Einstein, time and space change. But don't space and time exist independently of everything else? I would say no. Unlike energy and matter, space and time are only abstract quantities used to describe the world. You can't touch space or time. Ultimately they only become physical realities when they are measured,[1] and ultimately one always does this measuring with light or light-like waves. The measure of reality in space and on Earth is light; it doesn't just measure, it also defines space and time.

The thing that appeared before anything else in the Biblical creation story was light—with light came the first day. In the scientific creation story we tell today, light is also present at the beginning of time—and at the beginning of the universe there is a fireball, a big bang of light and matter.

But why is light so fundamental? After all, the universe is made up of matter, too—it's not all light! But when you delve further, you find that in fact, at the deepest level, everything is light and energy. Einstein's famous formula,

$$E = mc^2,$$

says that energy (E) equals mass (m) times light speed (c) squared. All mass is at the same time energy; all energy is also mass. In theory there is another variant of this equation, which is

$$E = h\nu,$$

where the Greek letter ν ("nu") stands for the frequency of light and h is the Planck constant, which translates light into energy. This is the simplest equation in quantum theory, whose founder was the German physicist Max Planck. In the smallest dimensions, for example at the atomic level, energy in the form of light can only ever be emitted or absorbed in certain energy units, so-called light quanta.

Light is therefore energy. The higher the frequency, the higher the energy. Matter and light are forms of energy and each can be transformed into the other.

To make things even more confusing, Einstein also found out that at high energy levels light sometimes acts like a particle. In such instances we speak of *photons*, wave packets inside of which light continues to oscillate while they whiz through space like little parcels of light.

So Newton and Maxwell were both right: light is a particle *and* a wave all at once—depending on what you're looking for. The answer is determined by the question! Today we know that this wave-particle dualism also applies to the smallest components of matter. Also, matter, in its most minuscule form, can sometimes act like a wave.

Even the forces of everyday life are communicated through light. What keeps atoms and molecules together are quantum physics and electromagnetic force—that is, the energy fields of which light also consists. In quantum theory, all these forces are mediated by the exchange of virtual light particles. When we touch each other or hit a nail with a hammer, this interaction is also communicated, at the tiniest level, by electromagnetic forces. Sound waves are produced when gas gets compressed and a pressure wave flies through the air. But when the air molecules in gas meet and knock against each other, they exchange the tiniest virtual light particles. Everything we sense, measure, perceive, or change is ultimately effected by properties of light. At the very smallest atomic level, all our senses depend on the exchange of light—not only sight, but also touch, smell, and taste. That's also why no information can ever reach us faster than at the speed of light.

Thus we always measure with light—and only what I can measure exists for me. In so far as this is true, a universe without light wouldn't exist at all. Space and time, matter and perception—they are all essentially nothing without light.[2]

The importance of measuring for the definition of reality is an insight that pervades the entirety of twentieth-century physics. It represents—even today—a radical revolution in thought, and is key to the theory of relativity and quantum physics alike, because in quantum physics the same thing applies: only when I *measure* something does it become reality. Everything else is interpretation, and interpretation, especially in quantum physics, is a matter of fierce debate,[3] as is the question of what measuring really means. Measuring always entails processes in the course of which particles exchange energy and light with one another. This thinking leads to completely new ways of seeing. In quantum physics a particle can, within a certain degree of probability, be everywhere at once—until it is measured. In the darkness of nothingness, everything is possible—until someone shines a light on it. Measuring means, for example, shining light on a quantum process. Since we are working in the smallest subatomic realm, the attempt to measure particles always also means to influence them, to change and pin them down, by means of photons. Measuring doesn't just define reality; it also changes it.

Erwin Schrödinger described this in his famous paradox. He imagined a cat in a closed shoebox together with a quantum-killer-device. So long as no one takes the lid off and looks inside, the cat is supposedly two things at once: dead *and* alive. Schrödinger's thought experiment is of course somewhat misleading, because the cat in the shoebox is not a single, isolated quantum object. Its particles are constantly exchanging virtual photons with each other and with the floor or the air. The cat is thus already being continually measured, or measuring itself—and this pins down its condition.[4] It doesn't just happen the moment we take the lid off. But of course it's just a thought experiment—and let's not even mention the fact that today no one would leave a poor cat to die in a box, not even hypothetically. You'd get into all kinds of trouble with animal rights activists, and rightly so!

A real cat is either dead *or* alive, not both. If, however, the cat were a lonely electron in empty space, far from other matter, then the statement would be logically correct. The electron wouldn't be here or there, but rather with a certain—sometimes vanishingly small—

probability would be everywhere and nowhere in space at once. Only when the electron cat is hit by a beam of light does this beam fix it in a particular place—and at precisely this point in time it is no longer located everywhere in space. Electrons can go through two doors at once, unless you install a photo sensor in one of the doorways that measures their passage—then they only go through one.

Thus we see once more the astounding and unique significance of light. Light creates reality—because it conveys information. Even space and time have their origins in light and matter. Space and time are abstract concepts that only become real by the action I take in keeping track of time or measuring out space. Without clocks there is no time; without a yardstick there is no space. The most elementary tool for measuring space-time is light. Only through its measurability does space take on physical characteristics, which we describe in models and representations.

If, however, light always travels at the same speed relative to *every* observer, then something else has to change for the observer, namely space and time. Albert Einstein was able to show this by means of a simple thought experiment, and from this he concluded that space and time are not absolute quantities that exist immutably, as Newton believed—they are merely relative. The only absolute is the speed of light.[5]

If, for example, a car is driving toward me, then the time in the interior of the car seems to flow differently than it does where I'm standing! This sounds bizarre, and it is, too, but the conclusion is unavoidable if you take the constant value of light speed seriously.

Let us consider some basic methods for measuring time. Mechanical wristwatches tick at a set frequency that is determined by the properties of a balance wheel. The regular ticking of the clock measures time second by second. We only need to count the number of ticks, then we know how much time has passed. Fortunately the minute and hour hands are kind enough to do the counting for us, so that we can just casually look down at the clockface.

The same principle applies for a digital clock—only here it is a crystal oscillating that determines the frequency. Ultimately, at the smallest atomic level, what is happening here is the transfer of energy via electromagnetic forces: virtual photons are exchanged. Even an hourglass depends on forces related to light when the sand molecules knock against one another and try to force their way through the narrow opening in the glass.

And so for simplicity's sake, let's build a pendulum clock in which it's not a weight that ticks back and forth but rather light that bounces between two mirrors. Set at a distance of 15 centimeters, the light would take about a nanosecond to go back and forth. Let's say we measure a billion light ticks per second. This equals a frequency of one gigahertz, or GHz for short. One hertz equals one cycle, or oscillation, per second. The unit was named after the physics professor Heinrich Hertz from Bonn, who was the first to produce and measure the electromagnetic waves predicted by Maxwell.

Now comes the crucial point: If I'm sitting with this light clock in the car, then to me the light appears to move vertically between the mirrors, up and down. If, however, a policeman is waiting by the side of the road and carefully watching the car as it drives past him at high speed, then it appears to him as if the light is moving up and down diagonally. The track the light makes resembles a zigzag pattern. You can picture this more easily if you imagine the light moving as slowly as the ant, the one in the car, which now crawls vertically up and down. The policeman sees it crawling up and at the same time moving to the side with the car—seen from his standpoint it is moving somewhat crookedly, and very fast.

The diagonal lines representing the movement of the ant and the light are naturally longer than perfectly vertical lines. In the same time interval, the ant and the light travel a greater distance to the policeman's eye. A naive observer might therefore conclude that the ant was traveling at super ant speed, and the light in the car was traveling faster than the speed of light. In the ant's case, that's true—but Albert Einstein and James Maxwell have forbidden light from doing precisely this kind of "speeding." A policeman with respect for the law should thus see the light moving at the same speed as the driver—even though from his perspective the light is traveling a greater distance.

How is this possible? The only answer: if from the policeman's perspective the distance the light travels is different, then the time there must also be different so that the speed can remain constant. Speed is distance over time, for example, kilometers per hour. If the stretch of track seems to change, then the time it takes to cover it must also change. Thus the policeman measures a time interval from outside that elapses the slightest bit more slowly inside the car.

This is called relativistic time dilation, and it flagrantly contradicts our intuition. We are used to the idea of speed being variable. If I have to take a detour in my car and I still want to arrive at the same time, I drive faster. Some people even recklessly risk a ticket for driving over the speed limit. With light that can't happen: it always goes at the same speed and simply changes the time, because light defines time. We all have to conform to time, but time itself conforms to light.

The whole thing sounds unbelievably abstract, just like the light clock in the car. In reality, though, no clock actually runs differently, right? In order to test this, the scientists Joseph Hafele and Richard Keating took a pair of flights around the world, first traveling in the direction of the Earth's rotation and then counter to it. They brought with them four highly precise cesium atomic clocks, whose times they planned to compare afterward with other atomic clocks on the ground. Would the clocks tick differently if they flew very fast and very far? The experiment was simple: the clocks they were able to borrow; the most expensive part of the experiment was the round-the-world boarding passes for the clocks, which they purchased under the name "Mr. Clock." These unusual passengers sat buckled up on their own seat. Airfare aside, this must be the all-time most reasonably priced experiment to test the theory of relativity.

And in fact, Hafele and Keating's experiment demonstrated that the clock that flew east—that is, with the Earth's rotation and with a slight difference in velocity relative to the clock on the ground—were 60 nanoseconds behind after the flight. The clocks on the flight traveling west—counter to the Earth's rotation and with a large difference in velocity relative to the clock on the ground—were a full 270 nanoseconds ahead of the clock in the laboratory.[6] The experiment was repeated several times afterward and impressively confirmed important aspects of the theory of relativity.

So we can't trust time, but then the distances that we measure aren't always the same either, because of course we measure distances with light, too. If the car speeds past the policeman at almost the speed of light, he can measure the length of the car with a stopwatch, calculating the length based on the car's speed and the time it takes it to pass. But if the driver of the car has two perfectly synchronized clocks, one at the front of the car, and one at the back, and measures the time it takes to pass the policeman, then the time interval he measures will be different, on account of the time dilation. The policeman measures a shorter time interval than the driver and as a result calculates a much shorter length than the person in the car. To the policeman the car seems much too small, while the driver enjoys his legroom.

And so we can't really trust space anymore either, not when things are moving—and this will have substantial consequences when gravity also comes into play.

MERCURY'S TIP-OFF: A NEW THEORY OF SPACE AND TIME

A few years ago a Dutch journalist called us up. He had some doubts about whether basic research was of any use to society and wanted to write an article on the subject. He led with the provocative question, "What good does it do us to take an exact measurement of Mercury's orbit?" I was flabbergasted. "Is this some kind a joke?" I fired back. "Is there a camera hidden somewhere?" I continued: "Mercury is the shining example of seemingly useless research that has gone on to fundamentally change our understanding of the physical world and made possible entirely new branches of industry." A Dutch company like TomTom, which sells navigational tools and software, owes its half billion euros in annual revenue to accurate astronomical measurements of Mercury's orbit and a patent clerk named Albert Einstein who made them possible. "Of all the things to make fun of, why would you pick on little Mercury?"

In the nineteenth century, after the laws of planetary orbit had been comprehended in all their beauty by Kepler and Newton, the planets became demystified. The sense of magic that had been attached to them was lost. After that, astrology, which had secretly kept scientific interest in the planets alive, was only pursued in esoteric circles, and today our solar system seems to us to be just a nice topic to learn about in elementary school. The problem seemed to have been solved, right? In fact, not everything was solved. A little problem emerged. Once again it all revolved, quite literally, around our solar system, and we saw how important it was to be able to take precise measurements.

Ever since Kepler, we've known that planets take an elliptical path around the sun. But that's not entirely true. Actually they're more like little flowers—rosettes, to be precise. These elliptical orbits aren't self-contained; every ellipse also rotates a little itself, with the result that when each planet reaches the point in its orbit closest to the sun, it's never in the exact same spot it was in the last time around. The effect is called perihelion precession. The perihelion, that is, the point in a planet's orbit closest to the sun, precesses—meaning, it moves—around the sun.

Planets don't just feel the gravity of the sun, but also the gravitational pull of other planets. With the help of Newton's classical theory of gravity, you can measure this effect quite precisely. But in practice it's not at all as simple as it seems, because in a system such as ours, every planet pulls on every other planet. If all the planets and the sun weighed the same, the whole system would fall apart on us. There would be times when two planets could tug on a third simultaneously and fling it out of the solar system. Our planets don't need to tug especially hard on one of their celestial colleagues to force it out of step; it's sufficient to give it a tweak at precisely the right moment.

It's like a child's swing that hangs on a long rope from a big cherry tree in the yard. A little push at the right time and the child starts to swing. If, however, you keep pushing continually at the right moment, then at some point the poor kid flies off the swing and into the neighbor's yard. In the same way there can be resonance between planets that take a uniform path around the sun, and that resonance can build and build.

When there are more than two swings or planets in play, it becomes impossible to keep track. It can be mathematically proven that the movement of even three objects in the same gravitational field cannot be determined exactly; the result is chaos in the truest sense of the word. Anyone who's been on a playground with small children will know this all too well. No wonder then that the *three-body problem* has kept mathematicians busy for many centuries and provided authors of romance novels with a never-ending wealth of material. The more bodies—planets or stars—orbiting around each other, the more chaotic things get. It can even be proven that it's fundamentally impossible to make any long-term prediction about the future paths of the orbits.

Chaos theory is by no means useless, however. It's true it can't predict the future, but it can identify the point in time when a system will become unpredictable. Our solar system also operates at the edge of chaos. For example, there's the chaos timescale, the so-called Lyapunov exponent, used to calculate the trajectory of planetary orbits for the next five to ten million years.[7] Extremely small changes can alter the future completely. Where exactly the Earth will be located in its orbit more than ten million years from now depends on where an ant coughs today.

When our solar system was forming, there was even greater chaos than there is today. In this primordial time our planetary system was filled with lots of little planetoids and minor planets. One by one, by means of the swing-set effect, they were flung this way and that, and sometimes even out of the solar system entirely. As a result of the mutual interactions between them, the large planets began to migrate either to the center or to the periphery. According to the Nice model developed by my colleague Alessandro Morbidelli and his coauthors, Uranus and Neptune might even have changed places. In our solar system, things weren't always the same as they are today, not by a long shot. The minor planets that managed to remain are brave survivors of a chaotic bullying phase that stretched over billions of years.

Incidentally, one of these leftover planetoids, which is listed as number 12654 in the register of the International Astronomical Union's Minor Planet Center, has borne the name Heinofalcke since 2019 and travels around the sun on a rather eccentric orbit. "That suits you," my old boss told me.[8] "It's probably just like me," I replied.

"It probably got bullied early in life and wouldn't let itself get thrown off track completely."

Chaos theory doesn't just apply to our solar system, but to many other systems as well, and it places a fundamental limit on our ability to know things in advance. This doesn't mean, however, that nothing at all is predictable. For example, we can have computers calculate statistically how minor planets will develop collectively over long periods of time. Sadly though, the resultant data can't tell us anything as specific as where exactly the asteroid Heinofalcke will be. I sincerely hope that its trajectory doesn't end up leading it to Earth at any point. It would be exceedingly unpleasant for me to hear on the news one day that Heinofalcke had just destroyed New York!

Thankfully though, a measure of calm has set in in our solar system by this point, and every planet seems to have found a more or less stable spot. There's no reason to fear that one of the planets could leave the solar system anytime in the conceivable future, and even little Mercury seems to be resilient enough to stand up to the larger planets' gravitational attacks—precisely because it has set up shop nice and close to the powerful sun.

Mathematically we view the reciprocal pull and tug between the planets as minor perturbations that we're able to calculate. The elliptical orbits gradually shift relative to one another, so that within certain bounds the perihelion precession of each can be exactly predicted. Within the span of the few centuries for which we have measurements, the chaotic proportion of the movement should be imperceptibly small. These calculations, based on perturbation theory, were a great success for celestial mechanics, and led in 1846 to the discovery of Neptune.[9]

Let's place ourselves back in the nineteenth century for a moment. Astronomers had explained the orbits of all the planets in detail. Well, not entirely. . . . One small indomitable planet still holds out against the astronomers. If you calculate the influences of all the other planets, then the axis of Mercury's elliptical orbit should rotate 5.32 arc seconds per year. In fact, however, it rotates 5.74 arc seconds per year—that's an annual discrepancy of 0.42 arc seconds.

Let's be clear about how very tiny this difference is. If you're dividing a birthday cake into twelve pieces, then each piece will equal a segment of 30 degrees. You can then divide each piece into 1,800 slices of one arc minute each. And each of these arc minutes can again be divided into 60 arc seconds. If we err by 0.4 arc seconds, then assuming the cake is 30 centimeters in diameter, we would have cut one piece too large, and that piece would be 300 times thinner than a human hair.

You really have to be a major pedant to keep nitpicking when the discrepancy is that small. But even a minimal difference adds up over time—and this kind of thing really gets under physicists' skin. If the measured results for Mercury don't align with the theory, then either the measurements were imprecise or the theory is off. Did someone overlook some small detail? If so, which, and where, and why?

For a long time, the culprit for this debacle was held to be a mysterious, undiscovered planet near the sun. Astronomers even had a name for it—they called it Vulcan, which of course would make its inhabitants Vulcans. Ultimately though, the Vulcans wound up in the realm of science fiction, and all because a young second-class patent clerk[10] had an entirely new, revolutionary idea.

SPACE IS JUST A BEDSHEET

At the beginning of the twentieth century Einstein placed our understanding of space and time on a whole new footing, embedding classical physics within his new theory of relativity.[11] Einstein was by no means the solitary genius type, toiling away at his major break-through, but rather a sociable bohemian and public intellectual.

In 1896 he began his studies at the Eidgenössische Technische Hochschule (Swiss Federal Institute of Technology) in Zurich, to-gether with Mileva Marić.[12] Einstein considered the young physi-cist his equal and, when it came to experimental physics, even his superior. They married when Albert got his first job. They would sit together for hours, conversing and reading philosophy books. Mileva and Albert also likely wrote their first articles together, but only Albert was named as the author.

Did Mileva choose to retreat into the background in order to im-prove Albert's career chances? Some think that by today's standards Mileva should have been named as a coauthor of the articles. "I need my wife; she solves all my mathematical problems," Albert is quoted as having said in the early phase of his career. It might be that Mile-va's primary focus was on their shared future. Asked once why her name didn't appear next to Einstein's on a patent application they had worked on together, she replied, punning on her married name, "the two of us are after all just *ein Stein*"—*one stone*. As a woman, it was most certainly more difficult, if not impossible, to establish yourself in the field of physics at that time. Historians today are still debating how large her scientific contribution to Einstein's ideas was, but it was definitely not insubstantial. The sources aren't clear. Einstein corresponded in writing with many physicists, but you can't search the archive for the ideas that were discussed at home at the kitchen table.

Einstein got his first job after college through the father of fel-low student Marcel Grossmann at the now-famous patent office in Bern. Then, in his annus mirabilis of 1905, Einstein published five groundbreaking articles. For his work on the nature of light, the "dis-covery of the law of the photoelectric effect," Einstein was awarded the 1921 Nobel Prize. Another essay posited that mass and energy were equivalent—the equation $E = mc^2$ might still today be the most well-known physics formula in the world. Finally, also in 1905, there came the article on the special theory of relativity, in which Einstein demonstrated that time and space are relative and change according to the relative speed of the observer. But Einstein wasn't done yet.

Even before Einstein's big day had come, his relativistic length contractions had already put the absolute nature of space in doubt. The second step began with Newton, a rotating bucket, and a spinning carousel. The British physicist had once pondered the curious qualities of a spinning bucket. Einstein pondered them further and determined that, because of length contraction, the relationship between the circumference and diameter of a rotating circle necessarily depends on the observer's location.

Let's picture a spinning carousel at a state fair with an axis in the middle and lots of kids riding brightly colored police cars, rockets, or wooden horses that are attached to a spinning circular floor. If a child waiting at the ticket booth measures the circumference and diameter of the round carousel with a tape measure, she'll find that circumference and diameter are proportionally related, and that the proportion is equivalent to the famous number π (pi).

Now, if a child sitting on a rocket ship on the carousel itself and spinning around in a circle were to measure the circumference with a tape measure, the child at rest, that is, the child by the ticket booth, would think that the circumference had grown smaller. Because of the relativistic length contraction, the length measured by the tape measure appears shorter to him. This apparent tape measure length depends on the direction of movement. The circumference of the carousel measured along the direction of movement seems shorter; the diameter, which is measured perpendicularly to the direction of movement, does not. The proportional relationship between circumference and diameter thus no longer equals π. This is surprising! Normal circles don't do such things: the circumference is always exactly π × d—that is, π multiplied by the diameter.

This is true for circles in a textbook, of course; there, the space where we inscribe the circle is flat. This changes, however, as soon as we look at a warped surface. For example, the children could draw a large circle in the middle of a taut bedsheet. If they hold the sheet at each end and lift it up together, then the two-dimensional plane starts to sag. The dimensions become warped, and the geometry of the circle changes: the circumference stays more or less the same, but the diameter becomes longer when you measure it along the surface of the bedsheet. The proportion of circumference to diameter in a warped space is no longer exactly π. It's important that this be a fitted sheet, because that's the only kind that really stretches out well!

It's possible to picture a warped bedsheet, but space is actually three-dimensional. This makes everything more complicated and harder to imagine. Can three dimensions also be warped? Our minds can't picture a warped three-dimensional space, but maybe we can describe it mathematically. Over time, Einstein came to understand that you actually even need a fourth dimension, because in the theory of relativity time also plays a crucial role.

The mathematical tools to describe the kind of spaces Einstein imagined had only just been developed in the nineteenth century. Warped—or curved—four-dimensional spaces are described with tensors: number tables that consist, for example, of four times four, or sixteen, entries. Every column or row stands for a spatial dimension, and with tensors you can do arithmetic as you would with plain numbers: you can add, multiply, subtract—but only if you know how it works.

Back then there were only a few experts working on this subject. They all had melodious names like Riemann, Ricci-Curbastro, Levi-

Civita, Christoffel, and Minkowski, and today you'll find them in textbooks on advanced mathematics. Except for Riemann, all of the above-named mathematicians were contemporaries of Einstein. These mathematics were brand new and too complex for Einstein. "I have developed an enormous respect for mathematics, whose more subtle aspects I have, up to now, in my ignorance, regarded as a luxury," he said.

No one does research entirely on their own. Thankfully, Einstein still had his old pal Marcel. "Grossmann, you have to help me or I'll go crazy," wrote Einstein—and this at a time when he was already a professor.[13]

Einstein and Grossmann were now posed with the challenge of working out the physics equations in such a way that they functioned in warped spaces. Inspired by Ernst Mach, a physicist and philosopher whom we know for having lent his name to the unit of supersonic speed, Einstein was convinced that natural laws must have the same form everywhere, whether it be at a picnic in the park, on a horse bouncing up and down in a spinning carousel, or in a rocket in space.

At first glance, taking on this challenge to find the universal applicability of physical laws seems obvious. But it allowed Einstein to bring the nature of space, time, and gravity together into one universally applicable theory, namely the general theory of relativity of 1915.

The crucial flash of inspiration came to Einstein when he was still working at the patent office in Bern. Whether he was the most dedicated of patent clerks, who's to say, but the job clearly left him a lot of time to think. This creative spark laid the foundation for the theory that today describes the expanding universe just as reliably as it does the gravitational force of black holes or the vibration of space-time caused by gravitational waves.

"This was the happiest thought of my life," Einstein said later. The crucial idea was that one can't fundamentally differentiate between gravity and any other normal accelerating force. "If a man jumps out the window with his eyes closed, he can't tell in that moment whether he's floating in space or in free fall—at least not until impact." This is more or less what Einstein was thinking.[14] Maybe at that same moment he was also daydreaming and thinking that if he closed the window, it would also be conceivable that he was flying through space inside a giant elevator. If the elevator continued to accelerate, then he would also be pressed back into his chair. How is he to know then if the force that keeps him in his chair comes from the gravitational pull of the Earth or the acceleration of the elevator? He can't tell the difference![15]

Locally, gravity and acceleration are indistinguishable. This principle is known today as Einstein's equivalence principle. It is a basic assumption and not proof. It is a principle, a dogma, that must be continually interpreted and tested through experiment.[16]

In reverse, the principle tells us that when one is sitting still in one's chair, one is also accelerating at the same time. That's exactly what it feels like! Even when we're sitting comfortably at rest, the same laws of relativity must apply as would in a fast-moving elevator or a powerfully accelerating rocket, which dictate that the space subject to accelerated motion seems to curve like a bedsheet.

But since Einstein can't decide whether he's sitting inside an elevator with the interior furnishings of a patent office or in a real patent office within the gravitational field of the Earth, then the Earth must also be able to warp space solely by means of the gravitational force of its mass. And in fact it turns out that gravitation warps not only space, but also time! Space and time must be considered together.

The conclusion is radical: gravitation is not a force; rather it expresses itself in the geometry of space-time. Since we still can't picture warped four-dimensional spaces, let's imagine space-time again as a taut bedsheet. If nothing or no one is lying on it, then it's even and level. If one places a bowling ball in the middle, then it causes a large indentation. If one places a billiard ball near the edge of the sheet, this creates a smaller indentation, and the billiard ball begins to roll toward the bowling ball. Actually both move toward each other: the billiard ball very quickly and the bowling ball just a little. The closer they get to each other, the faster they move, because the indentation gets ever steeper. The curvature of the sheet thus corresponds to the gravitational pull.

Next, if we flick a marble onto the bedsheet, then this will move in ever smaller elliptical orbits around the indentation caused by the bowling ball. On a flat sheet it would simply move straight; in a curved space, it follows a curved trajectory. Because of the friction on the surface of the bedsheet, the marble soon loses momentum, gets closer and closer to the heavy bowling ball, and finally drops into the funnel with the bowling ball at its bottom. Without friction, the marble would keep on moving and, just like the planets around the sun, follow its elliptical track for a long time, unhindered.

To go from that first idea to a convincing formulation of the general theory of relativity that now also encompassed gravitation, it took Einstein eight years—from 1907 to 1915—and many conversations, letters, and debates. At times he would think he had found his conclusive theory of gravitation, a concept, an "Entwurf," but then, again and again, he would toss the attempt out. Only toward the end of 1915 did he manage to set to paper a complete and coherent theory. Einstein was convinced that now he'd finally found the right answer.

He felt downright relieved once he had used his theory to calculate the perihelion precession of Mercury. Sure enough: his theory finally explained this minimal discrepancy that for so long no one could understand. The space-time indentations in the immeasurably large bedsheet around the sun made the orbit's circumference seem shorter; Mercury's ellipse was spun around a bit more quickly than previously expected. Einstein was "beside himself with joy for a few days." His heart was beating like a jackhammer. Newton was down—but not yet out.[17]

It's by no means sufficient for a new theory to be internally consistent and seem logical. Every theory has to prove itself both by experiment and in real life. It's a bit like how saints are chosen in the Catholic Church: After his or her death the candidate for sainthood must prove him or herself worthy by working not just one, but two miracles from beyond the grave. A single miracle is sufficient only for beatification.

The first miracle, which sufficed for the theory's beatification, was Einstein's having found the explanation for Mercury's strange perihelion precession. But the theory's canonization was still to come. Einstein's next miracle had to do, once more, with the qualities of light.

EXPEDITION INTO THE DARKNESS

Vision is of fundamental importance not only for us as people, but also for science. Our sense of sight enables us to find our bearings and convince ourselves of a fact's validity—or lack thereof. In astronomy especially, sight makes it possible to detect and experience things. Most of us have to see something first before we're convinced of it: as the saying goes, "seeing is believing," and rightly so.

Seeing isn't possible without light. But you also need the darkness to be able to better make out the essence of things. This was true during an expedition to what is probably the most famous solar eclipse in modern physics, on May 29, 1919. In making this research trip, the physicist Arthur Eddington wanted to test Einstein's general theory of relativity.[18] He wanted to demonstrate that starlight was deflected by the sun. What is especially worth noting here is that Eddington was British, and in conducting this experiment he would be helping the German Albert Einstein to global fame. This was by no means a commonplace thing in this year, a few months after the end of the First World War and after years of hostilities between the Entente powers and the German Reich. This expedition took extraordinary courage and is truly noteworthy in the history of physics.

According to the general theory of relativity, the mass of the sun warps the space-time surrounding it,[19] meaning that as a result, the sun deflects the light of the celestial bodies that lie behind it. It might sound crazy, but stars that, as seen from the Earth, were located near the sun would appear to be shifted slightly off to the side. Einstein's theory was mathematically flawless, but would it also hold up when subjected to an experiment? In order to find out the answer, astronomers needed a total solar eclipse, since during the day, when the sun is shining, we don't see any stars, while conversely, at night, we don't see the sun.

Eddington boarded a ship in 1919. He planned to measure the deflection of light that Einstein's theory predicted from the volcanic island of Príncipe off the West African coast. The British Astronomer Royal Frank Watson Dyson, who had organized the expedition together with Eddington, had sent a second team to Brazil. In May the sun was surrounded by the Hyades, a star cluster; the conditions were almost perfect. Eddington, already a proponent of Einstein's theory and a brilliant mathematician in his own right, was rubbing his hands together.

The sun was supposed to be covered by the moon for more than five minutes. But on the morning of the all-important day it was raining. Eddington grew nervous. Whether you're out at sea or be-hind the telescope, you're in God's hands, at least where the weather is concerned. Suddenly, just before the solar eclipse was set to be-gin, the sky opened up! The moon's umbra threw the observers into darkness. It was now or never: sixteen photographic plates were frantically taken, of which afterward only two contained usable data. The scientists had already taken a reference image without the sun before their trip. Meanwhile the intense sunlight had warped the metal casing of their colleagues' telescope in Brazil.

After returning home the scientists spent months analyzing the data. Then the breakthrough: the stars on the photographic plate were in fact shifted—by exactly two hundredths of a millimeter. Taking into account the discrepancy due to measuring error, this matched Einstein's mathematical prediction perfectly. They had gotten to the bottom of light's crooked ways!

"LIGHTS ALL ASKEW IN THE HEAVENS—Einstein Theory Tri-umphs," ran the headline in the New York Times. These measurements became the second miracle achieved by Einstein's big theory and made him a scientific superstar overnight. Still today the double expedition serves as a textbook example of the perfect interplay of theory and praxis. The nationality-transcending collaboration wasn't just a clear signal for the international scientific community after the First World War. After the turmoil of that war, it was a collective moment of joy and fascination that both friend and foe alike were able to share.

Strangely enough, Dyson himself had already photographed a perfectly identical solar eclipse in 1900, and the stars can even be seen on his photographic plates. When the data were analyzed at the time, however, they were looking for the mysterious planet Vulcan, and no one paid the slightly shifted stars any mind. Thus the answer's key component had been sitting in an archive for years—many years, in fact, before Einstein had even begun to formulate the special and general theories of relativity. It just goes to show that it's important to have a convincing theory and to ask the right questions!

The expedition was a triumphant success for Eddington, and even more so for Einstein. When Eddington presented his findings in November 1919 in London, the general theory of relativity didn't yet have many adherents. To older physicists, the upstart Einstein was already suspect, and many of them weren't even capable of understanding his ideas. Eddington was one of the few who could. When asked if it was true that only three people in the world had understood Einstein's theory, he is said to have answered: "Who's the third?"

The astronomical observations made Einstein's theory respectable, and we're still profiting from the results today in our everyday lives. Another prediction of the theory is that time also changes as a result of the curvature of space-time. Put simply, if light travels in a warped space, then it naturally has to travel a greater distance. But if the speed of light remains constant, then time has to expand as well. Light waves are pulled apart and oscillate slower. Time passes slower on Earth than it does in space.

When the first satellites of the United States' Global Positioning System (GPS) were shot into space in 1977, they were supposed to revolutionize navigation on Earth. On board they had extremely precise clocks whose time signals were sent back to the Earth's surface via radio. In planning the projects, physicists pointed out to the designers that according to Einstein the clocks would run faster in space because the Earth warps the space-time.

Somewhat reluctantly, the engineers built in a corrective mechanism, but they weren't quite buying what the physicists had told them. When they first launched the satellites into space, they had shut the corrective mechanism off. It quickly turned out that the clocks were in fact running 39 millionths of a second faster per day.[20] Since then the clocks are intentionally made to run a bit slower, using a corrective based on the general theory of relativity. On Earth they would be off, but as soon as they're in orbit, they're right, and we all silently reap the benefit.[21]

Today optical clocks are so precise that you don't even have to launch them into space to register the tiny differences in the curved space-time of the Earth. It's sufficient to lift these precision clocks 10 centimeters off the ground to cause them to register a speeding up of time relative to a control clock on the ground.[22]

The time corrective just outside the Earth's atmosphere is only minimal, but it is technologically significant nonetheless. All the effects we've mentioned are far more extreme when a lot more mass is compressed into a much smaller space and the warping of space becomes more pronounced. On the edge of black holes, time seems to stand still. In order to produce this warping effect, an enormously powerful force is necessary—the star forces.

THE MILKY WAY AND ITS STARS

THE HIDDEN LIFE OF STARS

To us humans, the stars in the sky always seem the same. But looks can be deceiving—they aren't the same. Stars change over very long periods of time. They lead their own unique lives; one can almost speak of each star having its own biography.

Stars are born and die. They are formed from dust and to dust they return. Like plants and animals on Earth, they are wrapped up in a continuous cycle of growth and decay. When they breathe their last breath and shoot their outer shells back into space, they assist in the birth of new stars. When a star is in its death throes, gas and dust are flung out into space and gather there in giant clouds that are then enriched by the ash of active stars. This chemical mix creates the perfect breeding ground for new stars and planets.

These interstellar gas and dust clouds, which can stretch tens or hundreds of light-years across, are probably among the most beautiful sights in the universe. A deep look into our Milky Way reveals how many interstellar clouds are there. Bizarre agglomerations of giant clouds can shine brightly, or they can drift in front of the light of the Milky Way and form dark shadows. With its powerful spiral arms, our Galaxy pushes them together like fresh snow under a snowplow. Looking at these formations through a telescope reveals them to be fantastic cosmic works of art.

Only 1,300 light-years away from us lies the Orion Nebula, one of the most beautiful clouds in our Galaxy. This glowing nebula is the only one that we can see with the naked eye under good conditions. Wreathed in a glowing veil of fog, the Orion Nebula is one giant delivery room for young and hot stars. The Orion Nebula glows mostly red and pink, with a scattering of blue—the effect is almost a bit gaudy. Its innermost center remains hidden to the human eye because the dust swallows all the optical light coming from the interior. Only at long wavelengths can astronomers penetrate this dust barrier and gain a sense of the center of such clouds. The infrared thermal radiation of hot gas, for example, penetrates through to the outside without much difficulty; the same goes for radio-frequency radiation. In the same way that x-rays shine through the human body, these waves can penetrate molecular clouds.

And just as the hot elements in gasses or on the surfaces of stars emit a unique bar code of a certain color of light, the molecules in dust clouds also emit a bar code.[1] High-frequency radiation is especially full of such lines. The wavelength of this light is only a few millimeters long, or even smaller. We are familiar with these waves in our day-to-day lives mainly from the modern body scanners at the airport.

On Earth we can measure the radiation from cosmic gas clouds. In the past 40 years, radio telescopes have been built all over the world to observe the behavior of such molecules in space. The largest interferometer in the northern hemisphere is located on the Plateau de Bure in the French Alps, 2,550 meters above sea level. There, the eleven silver 15-meter antennas of the IRAM NOEMA telescope glitter on the snow-covered mountainside. The largest facility of this kind in the world is the Atacama Large Millimeter Array (ALMA) in Chile on the Southern hemisphere. The ALMA Telescope consists of sixty-six dishes, most of them 12 meters in diameter. Operated jointly by European, American, and Japanese scientists, the telescope was built in the extremely dry and thin air 5,000 meters above sea level—the moist atmosphere at lower altitudes would absorb the tiny radio waves to too great an extent. It is radio telescopes like these that played a decisive role in producing the image of the black hole.

Let's return to space, to the birth of stars and gas nebulae. To us they seem like enchanted places in a far-off world. As if by magic, young stars are formed inside the cloud—though naturally there's no magic at play here; rather it's fascinating natural science. The gas nebulae consist overwhelmingly of hydrogen. This lightest of all elements is the crucial ingredient for the glow of the cosmos and the formation of stars. On Earth, small clouds of gas quickly disperse. In space, however, much larger quantities of gas gather together. Their gravity holds them together and they grow ever more dense. What exactly happens before the birth of a star is dictated by the Jeans criterion, named after the British astronomer James Jeans. In a cloud of this type, gravity and gas pressure are always in equilibrium. Jeans realized that various factors can upset this balance. If a certain point, known as the Jeans mass, is exceeded, the cloud contracts: it is now, as it were, pregnant and will give birth to new stars.

Occasionally all that's necessary is a small degree of compression and the cloud will start to become ever more dense under the influence of its own gravity. Bit by bit the temperature rises from −436 degrees to over 212 degrees Fahrenheit—the molecules in the cloud start to emit radiation and give off energy.

Once the gas reaches several thousand degrees Fahrenheit, the molecules and atoms start to break apart, the pressure drops, and just like that the whole structure implodes. The cloud collapses and breaks apart into small fragments. By cosmic standards at least, this happens very quickly; it takes less than 30,000 years for a little protostar to first glimpse the light of space. Already it is emitting warm, reddish light. Before it becomes a young star, it has to be patient and wait another 30 million years. During this time, as a result of the enormous pressure, the temperature rises to several million degrees Fahrenheit, until at some point nuclear fusion kicks in: now hydrogen is melted to become helium—just like in our sun. In the end a new star is born, like the thousands we see in the sky.

CLUMPS BECOME PLANETS

It's not only stars that form in these cosmic clouds. From our observational data today we can also deduce how entire planetary systems have formed and developed. When the clouds contract, the dust gathers in large disks slowly turning around the stellar embryo. The more matter contracts around the center, the faster it rotates.

We're all familiar with this effect from the pirouettes turned by figure skaters: when their arms are outstretched, they spin slowly in place. But when they pull their arms or legs in closer to their bodies, their speed of rotation increases. In sober and matter-of-fact terms, physics describes this process as follows: angular momentum is equal to the product of mass, distance, and speed and remains constant. If the distance decreases, the speed must increase. It's the same with the dust clouds in space that float around young stars or even completely envelop them. The more they contract, the faster they turn: now the disks of matter begin to form.

In essence what happens now is exactly the same as what happens in stellar formation: within the disk, small clumps start to form. I imagine it like making sauce from a packet in a saucepan: if you aren't paying close enough attention while the sauce is thickening and you don't stir fast enough, what you end up with isn't sauce but rather little clumps of powder spread around the pan. Only this time what is formed from the clumps of dust isn't stars, but rather planets. These protoplanets never get hot enough to trigger nuclear fusion inside their cores; their mass is too small and the pressure too low. The planets grow, suck up dust from their orbit, and plow grooves in the disks of dust around the young star. On images captured by the ALMA Telescope you can see such grooved disks around protostars; they look like a giant variation of the rings of Saturn.[2]

The rotational movement of the disks also explains how our planetary orbits came to be. All the planets were formed within a dusty primordial disk that spun around the sun. The slowly warming protostar that became the sun was the ice princess who gave birth to our planetary system.

Chunks of ice from the early phase of this formation can still be found on the outer edges of our solar system. These are the fluffy comets that form when water, rock, and dust gather together into dirty clumps of ice. Not every little clump in the spinning dust disk gets to become a little raw planet. Some become dwarf planets at best, like Pluto, or even smaller clumps of rock, like the planetoids and asteroids. They lack sufficient gravity to form nice round spheres.

Ultimately it was this celestial dust that brought the building blocks of life to Earth. Water and many organic molecules reached the Earth through this process and enriched it. All the elements we're made of were first baked inside stars, then frozen into molecules inside dust clouds, and finally came to us at the time of Earth's birth and early infancy. We humans are cosmic beings, and our bodies are made of stardust indeed.[3]

LIFE IN OUTER SPACE

When we see all this dust and all these planetary disks, we suddenly start asking ourselves: Couldn't there also be life somewhere else? Are we alone in outer space, or are there other life-forms out there? Even as a little kid I was asking myself this question—and almost every other person must have similar thoughts when they begin to grasp the dimensions of the universe.

When I began my studies in the midnineties, we knew of only a single planet outside our solar system. This planet revolved around, of all things, a dead star, the pulsar PSR 1257+12, and was discovered in 1992 by Polish astronomer Aleksander Wolszczan and his American colleague Dale Frail. The general assumption was that it wasn't the most habitable environment. In 1995, shortly after I received my PhD, Michel Mayor and his doctoral student Didier Queloz at the Observatoire de Haute-Provence, not far from Marseille, discovered another planet outside our solar system. Fifty light-years away, in the constellation Pegasus, Dimidium,[4] as it was later named, orbits the star Helvetios, which is actually quite similar to our sun. The two scientists received the Nobel Prize.

By this point we've found evidence of thousands of exoplanets, as the planets of other solar systems are called. But this is almost nothing compared to the number of planets there must be in the Milky Way alone. Statistically speaking it could be a hundred billion, maybe much more. But a clear sign of life has yet to be found. Still, it's very likely that we're not alone. By now more and more astronomers will dare to say this aloud and openly speculate about aliens.

Intelligent life could reveal itself through radio emission. Ten years ago I got some funny looks from colleagues in the Netherlands when a PhD student and I started combing through data from the LOFAR radio telescope looking for possible extraterrestrial signals.[5] The student would go on to work as a research fellow at the University of California, Berkeley, which received a full 100 million dollars to fund such projects from the Russian billionaire Yuri Milner. I'd sure like to have that kind of money for my own research. Even before Milner's gift, the astrophysicist Jill Tarter, who was immortalized in the movie *Contact*, had founded the SETI Institute in California with donated funds. *SETI* stands for "search for extraterrestrial intelligence."

We haven't yet found any intelligence in space—and a few of my colleagues would even say that there's none to be found on Earth either! All the same, the search for extraterrestrials has resulted in several technical advances that are useful for radio astronomy. SETI requires superb software and hardware that are able to process large amounts of data quickly. Astronomers now need the help of computer specialists such as Dan Werthimer, who started the SETI project at UC Berkeley. Werthimer was part of the famous Homebrew Computer Club and the milieu that included Microsoft founder Bill Gates and Apple founders Steve Jobs and Steve Wozniak. The latter three were also club members and went on to become filthily rich— but not Werthimer. Later we used Werthimer's fast computer processors to get a handle on the flood of data pouring into our telescopes.

In the end we owe the first image of a black hole not only to the submillimeter-wave telescopes that were built to illuminate the delivery room of stars and molecular clouds; we also owe it, in a small respect, to the once-so-eccentric-seeming search for extraterrestrial life.

Whether or not there actually is life somewhere outside of Earth, we won't know it until we've found it. For me it's a sober scientific question. No society or religion would collapse if we should discover extraterrestrial life. After a bit of excitement the world would go back to business as usual. Who we are depends above all on ourselves, not on any aliens that may be out there. All the potentially inhabitable planets are many light-years away, hundreds or maybe thousands, such that any communication would have to take place over generations. Instead of waiting for salvation from space, we should make sure that we keep our own planet in order and be mindful of how we treat one another.

DEAD STARS AND BLACK HOLES

DEATH IN THE HEAVENS:
HOW A STAR DIES

Stars are born and stars die, and in doing so they create space for new life, but also for black holes, which are formed from the death of stars—in space, absolutely everything is connected, and death holds its fascination and horror there, too.

A few years ago I was at a symposium in the US that was held in honor of the astronomer Miller Goss. From the sleepy town of Socorro, New Mexico, he had helmed the two largest and most successful radio interferometers in the United States: the Very Large Array (VLA) and the Very Long Baseline Array (VLBA). What was more important, however, was that he had lent support to many young scientists—myself among them. Colleagues from around the world had flown in to honor him. To mark the end of the symposium, he had organized a trip out to one of his favorite places. We drove to the famous Chaco Canyon, where sometime in the first millennium Native Americans had built impressive clay structures. On one end of these pueblos there was a small walled-in spot. This was where the stargazer once sat, claimed the bearded park ranger.

I imagined an old Native American who sat motionless in this spot every night, following the paths of the stars until the morning awoke in the darkness. It must always have been a sublime moment for him when the first rays of sunlight hit his body in the red light of dawn. Daybreak was an important daily ritual for him. Maybe a moment of relief at the continued existence of the Earth and of nature. And a silent symbol of the progression of time. Maybe it was also a moment of joy: life goes on. The light comes, the sun warms the ground, the birds begin to stir, and the sparse vegetation grows.

For the Ancestral Puebloans the canyon was a calendar. They could track the sunrise above a sharp cliff overhang and from this determine the exact day of the year. As a result of the Earth's rotation, the sun's starting point drifted a small bit southward in the fall and ever northward in the spring.

The old man who looked up at the stars here far before Christopher Columbus's arrival in the New World saw something else as well, however: because just under a thousand years ago, something unusual happened. A petroglyph very close by might possibly depict this extremely rare astronomical event, when a bright celestial object shone so brightly that it could be seen even in the daytime.

In the year 1054, people all over the world looked up at the sky in amazement. Some of them might have feared great catastrophe. The astronomers of the Song dynasty in ancient China made precise records of this cosmic sensation in their chronicles. There they reported on a "guest star" that shone as bright as Venus in the firmament. An Arab doctor even wrote of a new star.

In Europe as well, people may have marveled at a "bright disk" that dominated the sky in the afternoon, though confirmed accounts from Europe don't exist. What was so fascinating to people that they made records of this phenomenon all over the world?

It was a supernova, a gigantic stellar explosion.[1] It occurred at a distance of 6,000 light-years in our own Milky Way. The petroglyph in the canyon in which the old Puebloan once sat shows a half-moon that was painted in red on the yellow cliff face. Next to it a large star is visible. Round, with the usual rays—like a child might paint it. It's almost as large as the moon. This is how the Native American artists of the time depicted the supernova, the park ranger told us. Our group of astronomers wasn't entirely convinced. Experts debate whether the drawing depicts the famous supernova of 1054 or not.[2] But I also consider it unlikely that such an extraordinary event would go unnoticed.

You can imagine a star as being almost like a hot-air balloon. The heat in its core keeps it inflated. Once the fuel runs out, the gas cools down, the pressure drops, and the balloon deflates. Stars meet their end in the same fashion. Once their fuel is all burned up, they collapse. How and when they "die" depends on their mass. Light stars—and this comprises the vast majority of them—wear out after a long life and finally smolder out.

Our sun has an average life span. If it were to start collapsing in on itself, it could still activate its afterburner. In the heart of the star, the ash of nuclear fusion—a hot core of helium—has piled up. Under the high inherent pressure of an imploding star, the temperature starts to rise again, melts the helium into carbon, and releases its last reserves of energy; as a result, the "outer skin" starts to swell. Shortly before its demise, the sun will swell up, turn into a red giant, and swallow Mercury, Venus, and possibly even the Earth.

Stars with mass greater than our sun's eject gas and plasma out into space with their last breaths. Planetary nebulae are formed, which display wonderful shapes and colors when the dying star illuminates them from within. This spectacle takes place in a cosmic blink of an eye: after a few thousand years, these planetary nebulae fade. Their name is somewhat misleading, because the objects have nothing to do with planets, but they looked like distant planets made of gas when they were discovered in the eighteenth century with the telescopes of the day.

At their center is the compressed ash of nuclear fusion, on which the entire weight of the burned-out star weighs. This pressure becomes so great that the atoms get more and more scrunched together until they don't have any room left and are standing shoulder to shoulder, so to speak. Electron pressure then prevents the further collapse of the star. The electrons that revolve around the individual atomic nuclei in the core of the star are called fermions. Fermions are the loners of physics; a fermion can't share its bed with any other fermion. When things get too cramped, fermions counteract the pressure of gravity and in so doing prevent the complete collapse of the burned-out core.

If the outer regions of the star have been shed, all that's left over is the small, tightly packed, brightly glowing core of carbon—a white dwarf, as large as the Earth but as heavy as the sun. A teaspoon of the matter that makes up a white dwarf, which our sun will become in a few billion years, weighs about nine tons, as much as a delivery truck. The dwarf's surface remains very hot and radiates thermal energy out into space for a very long time before the dead star finally turns into a cold, perfectly round carbon crystal—a giant diamond in space.

Different quantum mechanical effects play a role in this process and were calculated by the Indian physicist Subrahmanyan Chandrasekhar. In 1930, at the age of just nineteen, he traveled to England by ship to continue at Cambridge the course of studies in physics that he'd begun in India. During the passage he had a lot of time on his hands, so he set out to calculate the maximum mass that a white dwarf can reach and arrived at a value of 1.44 solar masses.

What would happen, however, if a star should be much larger and heavier than our sun and the pressure should rise to the point where it's literally unbearable? A star that is more than eight times heavier than the sun will ignite more afterburners to prevent its collapse. Onion-like, it burns off layer by layer—the core of the giant sun is burning itself up. The closer a shell to the center, the hotter it gets, burning the ashes of the previous layer surrounding it to form ever more massive atomic nuclei, and with each layer another reserve of energy is unleashed. Hydrogen becomes helium, helium becomes carbon, carbon and helium become oxygen, oxygen becomes silicon, and silicon becomes iron. Every incendiary process goes more quickly than the one before. It takes a million years for helium to be burned into carbon, but the fusion of an entire quantity of silicon to form iron takes only a few days.

And then that's it! In terms of energy, iron has the most compact atomic nucleus in all of nature. If the pressure is great enough that iron melts to form still more new elements, this process will not generate more new energy but rather will require energy. Suddenly the trick of simply increasing the pressure and squeezing more and more energy out of the atoms doesn't work anymore. Just like that, atomic cooling, rather than heating, enters the process; the pressure starts to decrease rather than increase. The last teetering support propping up the enfeebled old star gets knocked away and it collapses into its death. Within a few minutes the core implodes: the dying star can no longer withstand its own gravity.

The internal pressure in the stellar cadaver increases to such an unimaginable extent that even the tightly packed atoms are crushed because the core of this kind of star is heavier than the maximum limit for white dwarfs calculated by Chandrasekhar. But there is still one last stop before the irrevocable, eternal collapse. The otherwise contact-resistant electrons escape into the atomic nuclei and fuse together with the protons to form neutrons: the outer shell of the atom disappears into the nucleus and what remains is 10,000 times smaller than before.

If you imagine an atom with an electron shell as large as Rhein-Energie Stadium, where my favorite but not always successful soccer team 1. FC Köln plays, then the nucleus is only as big as a nickel placed on the center mark. The matter we know, which consists of atoms, normally contains a lot of empty space. If a star's atoms become pure neutrons, then it shrinks down and becomes a neutron star. The collapse is like squeezing an entire stadium into a little coin. In a neutron star, more than one and a half times the mass of the entire sun is collected in a sphere with a diameter of just 24 kilometers. The density is unbelievably high. Five milliliters of neutron star matter would weigh 2.5 billion tons—that's 8,000 times the mass of the cathedral in Cologne in a single teaspoon.

For a long time, neutron stars sounded like wild speculation, and they were just that until Jocelyn Bell and her doctoral adviser Antony Hewish wrote history and discovered a strange radio signal at the Mullard Radio Astronomy Observatory in Cambridge on November 28, 1967. Many short pulses reached the Earth at precise intervals. The object thus came to be known as a pulsar—it was as if a clock were ticking in space. At first the two researchers were a bit baffled by this precision and named the radio object, half-jokingly, "LGM," for "Little Green Men." Soon it became clear that they had discovered an extremely small, unusually heavy object that was spinning on its own axis at immense speed. It was in fact a neutron star—a dead star as heavy as the sun and as large as the Nördlinger Ries, an old asteroid impact crater in Bavaria. Not every neutron star becomes a pulsar, but every pulsar is a neutron star. Like a cosmic lighthouse, it beams its radio emission out into space as two rays of light that reach us at regular intervals and cause flashes of radio lightning in the sky. Because the object is so stable and massive, it functions like a giant balance wheel. It ticks more precisely than any atomic clock. On account of their extraordinary stability and consistency, we can use pulsars to carry out many tests of the theory of relativity.[3] A famous example is the double pulsar PSR J0737–3039,[4] which is actually two pulsars that orbit each other. The precession of the elliptical orbit—the same phenomenon we saw with Mercury, which made Einstein's heart start racing when he figured it out—is here 10,000 times more rapid and has been accurately calculated to five decimal places.

The emergence of neutron stars is a spectacular event that is seen to occur in stars with more than eight times the mass of the sun. A super-sun of this sort dies in a more impressive fashion than our own star will. The super-sun's burnout becomes a galactic firework. Under the pressure of its collapsing mass, the core suddenly gives birth to the new neutron star, but the rest of the star implodes at supersonic speed. Electrons and protons suddenly fuse together in the nucleus and release large amounts of tiny neutrinos, which deposit even more energy in the outer shell of the star. Now a catastrophic shock wave travels outward through the entire star and finally tears it apart. Astronomers call this kind of galactic bang a supernova. It flashes up in space and is quite an impressive thing to watch. And it is just such a spectacle that the Native Americans in Chaco Canyon and many other stargazers all over the world could have marveled at.

Let's try to imagine a supernova: In just a fraction of a second, this galactic explosion releases more energy than the sun has produced in its entire life span. Even then it takes a few weeks till all the light has torn through the expanding outer shell of the star. As a result one can sometimes observe a supernova for months. In the extreme temperatures and pressure conditions that result, more new elements are created that are heavier than iron. Cobalt, nickel, copper, and zinc are flung out into space in rubble clouds consisting of gas that is millions of degrees Fahrenheit and still smoldering.

These interstellar shock waves travel through space at tens of thousands of kilometers per second. They spread out in a spherical shape and are massive cosmic particle accelerators. Some atomic nuclei are propelled almost at the speed of light and drift through the Milky Way along the turbulent magnetic field in interstellar space. A vanishingly small portion of them rain down with great energy on Earth and are part of what we know as cosmic rays.

We're still seeing these shock waves today. In 2009 one of my former students[5] discovered a new radio source in our neighbor galaxy M82. We saw a bright radio ring that expanded at a rate of 12,000 kilometers per second over the course of a few months.[6] Based on the speed and size, we were able to conclude that the star had exploded a year prior. We had discovered the supernova 2008iz. It lay behind an expansive dust cloud, which is why it had stayed hidden to all other telescopes. It's extremely exciting to discover and experience for yourself, live, the kind of cosmic drama that you otherwise know only from either science-fiction movies or dry academic literature.

Today we can still see the remnants of the bright supernova of 1054. It left behind the spectacular Crab Nebula. This object in the Perseus Arm of our Milky Way looks like a colorful puff of smoke and proves that the ancient chronicles weren't fairy tales.

There are only an estimated twenty supernovas per millennium in our Milky Way. One of them took Tycho Brahe and his sister Sophie by surprise on November 11, 1572. Because they took the event for the formation of a new star, they also coined the name—*stella nova,* meaning "new star." In 1604 Johannes Kepler also described a supernova. The lack of any parallax showed that this light didn't come from our atmosphere but rather had to come from at least beyond the moon. This supernova was yet another nail in the coffin for the Aristotelian model of the universe, which assumed the celestial spheres of fixed stars to be immutable.

Today astronomers are discovering new supernovas all the time—they're all the way out there, in other galaxies. But any day now a new supernova from our own Milky Way could appear in the sky and we could see it with the naked eye. Actually, it's about time for the next supernova, but it could also be another hundred years.

There's no danger to mankind even from very close supernovas. In the grand scheme of things, we even have these stellar explosions to thank for the formation of our planets and of life on Earth. This is because, in its last phase of life, a star produces important elements in ever more rapid cycles. By means of the supernova, they are then flung out into space, where they collect in massive dust clouds from which, in subsequent stellar generations, new stars and planets can be formed. This is the origin of all the important elements on Earth. Thus, without the death of stars, there would never have been life—or even the pretty red color chosen for the Golden Gate Bridge. It contains iron oxide, the iron of which was ultimately first forged by a supernova explosion. So we have a lot to thank dying stars for.

A BLACK HOLE FORMS

There are stars that are too massive even to become neutron stars. Think of this like the particularly stable chair in the living room that is reserved for your extremely overweight Uncle Alfred. Ever since the time when he sat down on a cheap plastic folding chair and broke it, he has always been given this massive wooden chair to sit on. Better safe than sorry. But even the most stable chair has its limits. If Uncle Alfred brings his circus elephant along and lets him sit on the wooden chair, then it, too, would break.

In astrophysics, white dwarfs are the cheap plastic chairs, and neutron stars the stable wooden chairs. They can withstand a lot, but not everything, because there are some real elephants among the stars. We owe this insight to none other than the father of the American atomic bomb, Robert Oppenheimer, along with his colleagues and students. They proved shortly before the Second World War that even neutron stars have a maximum mass limit, just as Chandrasekhar showed that white dwarfs don't have limitless capacity.[7] According to today's calculations, this maximum mass for neutron stars is a bit more than two to three times the mass of the sun.

The stellar elephants in the universe are stars that weigh more than twenty-five times the sun. When these stars explode, the majority of their mass flies out into space, while in the core first a white dwarf is formed and then a neutron star. Inside the core, more and more matter plunges toward the center, so that at some point even the neutron star collapses. Once this happens, there's no resistance left. We don't know of any force that can withstand the weight of such a heavy star—the collapse is inevitable. The star continually crumples into itself and becomes smaller and smaller, until at some point all the mass is concentrated in a single infinitely dense point. Now one of the most remarkable objects in the universe has formed: a black hole. Though of course in Oppenheimer's time it wasn't called that yet.

Albert Einstein himself was frightened by the thought of such an object. Just a few months after Einstein had developed the theory of relativity, the German astronomer Karl Schwarzschild had already worked out the space-time structure for mass concentrated into a single point, and the consequences were disturbing in the extreme.

Schwarzschild was a pioneer of modern astrophysics. He was director of the Astrophysical Observatory in Potsdam when the First World War broke out in 1914. Unlike Arthur Eddington, the pacifist and admirer of Einstein, Schwarzschild, the son of an upper-class Jewish family, chose to serve his country and volunteered for the German artillery. A tragic decision. Two years later he took ill at the front and died.

Nevertheless, Schwarzschild succeeded in writing two world-class scientific papers in the midst of the war.[8] In one he calculated the curvature of space-time around a point mass. In so doing Schwarzschild became the first to calculate a clear solution for the equations of the general theory of relativity in a concrete case,[9] and he proudly sent it to the surprised Einstein. "I would not have expected that the exact solution to the problem could be formulated so simply," Einstein wrote back, and presented the results to the Prussian Academy of Sciences at its next meeting.[10]

In Schwarzschild's solution,[11] all mass is concentrated in a single point; in this spot, however, space itself seems to expand infinitely in one direction, and the degree of curvature of space becomes infinitely large. Suddenly in this limited part of space there is an infinite amount of room. The equations show a singularity—one of those points where equations just about explode, where they vault into the infinite and everything comes to a stop. We physicists learn that singularities aren't reality; rather, they demonstrate that something is still missing in the equations. To Einstein, then, it was clear: point masses don't exist. This is a purely mathematical gimmick, albeit an interesting one.

What was unsettling for Einstein and other scientists, however, was that far outside the central singularity something curious was happening in the equations, namely at a distance of

$$R_S = \frac{2GM}{c^2}$$

This distance is known today as the Schwarzschild radius. M stands for the mass of the object, $c = 299{,}792.458$ km/s, and $G = 6.6743 \times 10^{11}$ m^3/kg/s^2, or light speed and the gravitational constant, respectively.

Something was wrong out there. The equations were acting crazy. Once you reached the Schwarzschild radius, time seemed to stand still. Once you were within the radius, however, you were no longer traveling through space, but rather, in a sense, through time.

In normal life I can sit quietly on a park bench—I'm sitting at a fixed point in space, but time continues to pass. Inside the Schwarzschild radius I remain stuck in time, but space pulls me irresistibly inward toward the central singularity. With every attempt I make to move outward, I only get closer to the center.

This is bizarre. There appears to be no possibility of ever leaving this space again, of crossing the Schwarzschild radius from within. Once something finds itself beyond the Schwarzschild radius, there's no escape—not for matter, not for light, and thus not for information or energy either. It took a long time before anyone knew what was actually happening there. Without knowing it, Schwarzschild, writing from a dismal trench in the middle of the First World War, had described a black hole.

But really it was clear even before then that something had to go awry when you got near a point mass. Wasn't it already apparent in Kepler and Newton's simple theory concerning planetary motion? The closer you get to the sun, the faster you move around it. If the sun were infinitely small, a planet circling it in a tight orbit with a radius of just three kilometers would have to travel at the speed of light; farther in than that, it would have to travel even faster than light. But of course that's not possible!

The gravitational force becomes too great as well. The more mass is concentrated in the same space, the greater the gravitational pull, and thus the more difficult it is to break free of that pull. If you want to escape the Earth's gravitational pull you have to fire a rocket into space at a speed of 11.2 kilometers per second. From the surface of the heavier sun, you would need to travel 617 kilometers per second.

If you were to compress the sun even further, the escape velocity on the surface would continue to increase until at some point you would have to fly faster than the speed of light. But then, in Newton's theory, not even light would be able to escape; it would fall hopelessly back onto the star. In Einstein's theory, however, if you're at the edge of a black hole and traveling at the speed of light, you can't even move forward anymore!

Way back in 1783 it had occurred to Reverend John Michell, without any notion of the theory of relativity, that something like this would have to occur in nature if a star had enormous gravity and the escape velocity were greater than the speed of light. A "dark star" of this kind would have to be invisible, because no light could escape it, even though it existed and was located at a specific coordinate in space.

In Einstein's theory the space around a black hole is like a rushing river[12] that ends in a waterfall at the Schwarzschild radius. Light is like a swimmer in this space river. Far off from the edge, it's still able to swim against the current. Closer to the waterfall the current grows stronger, and it has to swim faster and faster. At some point, though, not even a world-champion swimmer can escape the rushing current; the swimmer is swept away. Once you fall over the edge of the cliff, it's too late. No one can swim up a waterfall. The exact same thing happens at the Schwarzschild radius. It's a point of no return. There, not even your cries will get out. Even light, along with space, is pulled into the depths.

In 1956 the physicist Wolfgang Rindler coined the term "event horizon" for this "uncanny boundary." The event horizon can be neither touched nor felt—it's only a certain margin in empty space, a mathematical definition, and yet a dividing line.

If one calculates the Schwarzschild radius for the sun, one gets a value of 3 kilometers, for the Earth one of 0.9 centimeters, and for a person like me a value one hundred-billionth the size of an atom's nucleus.

Einstein was convinced that the area inside the Schwarzschild radius was unphysical—pure fantasy, pure mathematics. Nature would surely prevent such objects from ever forming. In 1939, he published an essay in which he tried to prove with the help of his theory of relativity that such "dark stars" didn't exist. He closed triumphantly with these words: "The essential result of this investigation is a clear understanding as to why the 'Schwarzschild singularities' do not exist in physical reality." This meant nothing other than: black holes don't exist.[13]

But Einstein's article missed the mark. Indeed, at almost the same time, Oppenheimer and his colleagues were demonstrating that stars could most certainly collapse into a single point.[14] If they were massive enough, there was no preventing the collapse from taking place.

But here the remarkable qualities of the theory of relativity were on display once again. What one would see during the collapse depended to an extreme degree on one's location. An observer following the collapse closely with a telescope would see the star implode and disappear into a black hole. An event horizon would emerge and everything that approached it appeared ever weaker and slower to her. Every light wave would be stretched out infinitely wide and be no longer measurable by her once it tried to escape from the edge. Time would become viscous and syrup-like and finally seem to come to a standstill. If we imagine light waves as the timer in a clock mechanism, then they, like space, are stretched farther and farther out. The clock ticks more and more slowly until at some point it stops ticking altogether.

Meanwhile, for the careless observer who remains seated on the surface of the collapsing star, nothing particularly special happens— except of course that he plummets to certain death. He falls into the core of the star along with all the other particles. As he passes the event horizon, he doesn't notice anything out of the ordinary—he doesn't even notice he's passed it. The black hole always appears before him as a big black spot—even inside the black hole. His time also continues to pass normally, until eventually, within a fraction of a millisecond, he is squashed into a single point in the core of the star. His light falls in with him. In the case of a stellar black hole, however, the fun here is extremely limited. Because the feet of our reckless observer are closer to the center of mass than his head, they'll also be attracted more strongly than his head—he'll be pulled apart and stretched out like a spaghetti noodle.

Physicists have fun imagining such scenarios, even if this fun isn't shared by everyone! For a long time these objects were called "frozen stars," because time comes to a standstill at their edges. But that's not exactly the case. Strictly speaking, the halting of time only applies to the edge of an eternally static black hole. If it grows, because it is devouring matter, its event horizon also grows and steamrolls, so to speak, the "frozen" matter that falls into it.

The term *black hole* appeared for the first time in 1964 in an article by the journalist Ann Ewing;[15] it was then taken up by John Archibald Wheeler at a conference, and thus became established. Since then black holes have fascinated both laypeople and experts alike. Words are important in physics, too, and Americans know a thing or two about marketing. Nobody today would buy a book about the first image of a "gravitationally fully collapsed object."

But black holes can also rotate. The mathematician Roy Kerr from New Zealand discovered a mathematical solution for such an object in 1963, which describes the space-time around such a body.[16] If rotating matter falls into a black hole, the angular momentum is conserved. The black hole causes space to rotate along with it, in the same way a whirlpool causes water to rotate. And just as a boat is seized by a whirlpool and pulled into the depths, so does the rotating space force matter and even light within a certain distance to rotate with it. Conversely, it is possible, in theory, to extract rotational energy from a black hole with the help of incident magnetic fields in the whirlpool area.[17] The singularity at the center of rotating black holes is a ring with insane qualities. Mathematically one can circle it, setting off at one point in time and arriving at the exact same moment.

Black holes are only formed by very large stars that don't live long, maybe just a few million years. Shortly after a giant star is formed, it explodes again. Wherever young stars are formed, stellar black holes are also created soon thereafter. There are an estimated one hundred million of them by this point in our Milky Way. They are thousands of light-years away and too small for us to be able to capture images of them. Sometimes one sees them shining in the sky as bright sources of x-rays, namely right at the moment when they suck up matter from a neighboring star that is orbiting them. This kind of pair is referred to as an x-ray binary. In reality it's a star and a star cadaver that are orbiting each other. The black hole zombie eats its partner bit by bit.

IN THE HEART OF THE MILKY WAY

It's June 2016 and I'm sitting on a sweeping, flat-topped mountain, Gamsberg mountain in Namibia, where we'd like to build a new radio telescope.[18] As of now there are just a few huts here—we don't have the money, but I can still look off into the distance, surrounded by an overwhelming panorama. Below me the stony and varicolored desert stretches to the horizon in every direction, above me the setting sun swathes the almost cloudless sky in a deep red. My eyes are dazzled by the slowly vanishing play of colors between sand and sun. Is there a single more beautiful moment than this one? My gaze into the heavens is never purely objective; it is always mixed with fascination.

In the clear, dry nights of southern Africa—far away from the nearest town—the starry sky soars above me like a great painted dome. The glow of the Milky Way in all its splendor is set off from the darkness of space, a bright band: 100,000 light-years in a single sight. Countless stars are woven together in a glowing veil that spreads out over the entire sky. Black spots give it a plasticity that I am unfamiliar with, being used to seeing it from the Earth's northern hemisphere. They are interstellar dust clouds, breeding grounds for new stars, planets, and black holes—I can spot them with the naked eye. Straight up, almost directly above me, is the core of the Milky Way. There, hidden somewhere at its center, is "my" black hole. Under the clear starry sky it seems close enough to touch. But where exactly it is I can only guess, because the dark dust clouds of our home Galaxy block my view into its heart. As beautiful as the Milky Way is, it makes it hard for us to fully comprehend it—precisely because we are part of it. We aren't just observers; we're also inhabitants of this island in the cosmos.

Next to the moon, the Milky Way is the most visible formation in the night sky. It shines so bright and clear that, according to legend, it led the apostle Saint James the Great to Santiago de Compostela. Today I use GPS when I walk the "Camino," but dung beetles at least still use the Milky Way to orient themselves when rolling their balls of dung away from a dung heap.[19] This white strip must have lent inspiration to the thoughts and feelings of the very first hunter-gatherers.

The Milky Way got its name in antiquity. According to Greek myth, Zeus placed his son Heracles at the breast of his wife Hera when she was asleep. But the goddess was wakened by the strong pull of the suckling child and pushed Heracles away. When she did so a bit of breast milk was scattered across the vault of heaven—the Milky Way was born. In Greek it is called *Galaxias*, which is where our term Galaxy comes from. The Milky Way is made up of hundreds of billions of stars. Other Milky Ways in space are called galaxies. The naturalist and explorer Alexander von Humboldt called them *Welteninseln*—literally "islands of worlds"; commonly translated as "island universes"—which for me is an even more beautiful name for them.

Democritus, the Greek philosopher, postulated in the fifth century BC that the light of the Milky Way could only come from the combined glow of several individual stars. Almost 2,000 years later, Galilei, observing the stellar abundance of our Milky Way with his telescope, saw that Democritus was right. Immanuel Kant wrote in the seventeenth century that the Milky Way must be arranged like a disk and that its stars must be configured roughly along the same plane.

Around that same time, the French astronomer Charles Messier went comet hunting from the Hôtel de Cluny in the center of Paris, which today houses the Musée National du Moyen Âge. He discovered many strange cloudy spots in the sky that were clearly not comets and didn't move. What these clouds were, Messier couldn't say—but he documented and numbered them. He compiled 110 of these diffuse apparitions in the catalog that now bears his name.

Today, amateur astronomers still love tracking these Messier objects down. They're abbreviated with an *M* for Messier followed by their catalog number. M1 is the Crab Nebula, which was formed from the supernova of 1054. The Hercules Globular Cluster M13 is the brightest globular cluster in the northern hemisphere, lying at a distance of 22,000 light-years. Hundreds of thousands of old stars circle one another in an orbit with a diameter of 150 light-years. M42 is the Orion Nebula, in which stars are born.

All these objects are part of our Milky Way. It is full of wonderful structures and clusters. But not all objects in the Messier catalog are part of our island universe. M31 is the Andromeda Galaxy, in earlier times called the Andromeda Nebula—the twin galaxy to our Milky Way, located right next door to us at a distance of 2.5 million light-years. And M87, also called Virgo A, in the Virgo constellation, is a monster galaxy with a few trillion stars and our mighty black hole in the middle. In his time, Messier knew none of this. For him it was just important to produce a usable list so that no one would mistakenly confuse these cloudy spots with comets.

Toward the end of the eighteenth century William Herschel gave a sense of the true dimensions of the Milky Way. Herschel was an amateur astronomer who earned his living as a musician and composed symphonies and fugues. His true passion lay with the stars, however, which he observed together with his sister Caroline. Caroline, a singer, was likewise a gifted astronomer.

Although he was an autodidact, the German-born Herschel had earned a reputation as one of the best makers of reflecting telescopes. He even poured the mirrors himself, which could be larger than a meter in diameter. Herschel supplied telescopes to scientists and noblemen all over Europe, and even sent one instrument to China. Most of all he liked to observe and explore the starry sky himself using his largest construction, a 1.2-meter telescope that was supported by a giant wooden frame and required a system of pulleys and hoists to move.

Herschel was the son of a military musician and moved to England when his father was ordered there. Here the Hannover-born Herschel siblings counted the stars and expanded the catalog begun by Charles Messier. The Herschels discovered that some of the clouds described by Messier were actually made up of individual stars. In 1785 the two published an image of the Milky Way with 50,000 stars. The vaguely oval-shaped image doesn't have much relation to reality, true, but that's more a fault of the method the siblings employed than the data. In the Herschels' model our sun is still more or less in the center of the Milky Way. A misconception, as we know today.

At the start of the twentieth century the research was already showing an astonishingly accurate picture of the galaxy. Astronomers assumed a flat disk shape with a diameter of about 100,000 light-years. Its vertical extent was about 4,000 light-years. Nevertheless, most scientists still supposed the sun to be in the center.

The next step was taken in the early years of the twentieth century by the Dutchman Jacobus Kapteyn, who at just twenty-seven years old was already a professor of astronomy at Groningen. He realized that all the stars moved around a common center. Kapteyn published his dynamic Milky Way model in 1922, although he, too, was off in one crucial respect—thankfully, because in his model our solar system was still to be found very close to the center point, and based on what we know now that would have put us in the vicinity of a giant black hole.

The US astronomer Harlow Shapley corrected this. He did his research at the Mount Wilson Observatory, where he worked with a giant telescope. Shapley deduced the size of the Milky Way by measuring globular clusters and determining their distance to Earth.

This was only possible because his countrywoman Henrietta Swan Leavitt had discovered in 1912 how one could determine the distance of certain stars, in this case Cepheid variables, from regular and periodic fluctuations in their luminosity. Leavitt, like Annie Jump Cannon, belonged to a generation of passionate and indefatigable female astronomers whose achievements were not always given due recognition. Today, at the very least, Cannon and Leavitt have lunar craters named after them.

Determining the position of these globular clusters showed Shapley that they weren't centered around the sun. This meant that the spiral arms of the Galaxy couldn't possibly revolve around our planetary system, which in turn meant that the Galactic Center of the Milky Way had to be much farther away from us than Kapteyn had posited. Shapley estimated the position of our solar system to be about 65,000 light-years away from the center. Later he corrected the distance to about 35,000 light-years. This made Shapley the Copernicus of our home Galaxy. The German-Polish canon had once taken the Earth from the center of our planetary system and shipped it off to a more distant orbit; now Shapley was banishing the sun and its planets from the central hub and fulcrum of our Milky Way out to its periphery.

Shapley believed the dimensions of the Milky Way to be considerably greater than what had been posited up to that point. According to his estimates, its diameter had a length of 300,000 light-years. All those nebulae out there were part of our Galaxy, or so he assumed. Therefore there could only be one Galaxy—ours. The whole universe, so his thinking went, consisted only of our Milky Way.

By taking this view, Shapley got himself involved in a legendary discussion. On April 26, 1920, the "Great Debate," as this clash was later dubbed, took place at the National Museum of Natural History in Washington, DC. Two schools of astronomy faced off against each other: in one corner there was Shapley, arguing for his giant galaxy with the sun distant from the center; in the other, his critic, Heber Curtis, representing the island universes theory. Curtis believed that the Milky Way was only one of many galaxies, and that the spiral nebulae each comprised their own independent stellar system. In his model, though, our solar system occupied the central position inside the Milky Way.

At the meeting both scientists had given lectures during the day presenting their respective theories. The showdown came in the evening, at an open discussion. Neither of the two would give the other an inch. Curtis, who over the course of his career had led several observatories and undertaken almost a dozen expeditions to view the solar eclipse, was certain that Shapley had botched his measurements. Both argued their positions vehemently, but there was no clear winner that evening. By the end of it, Shapley had probably won a few more listeners over to his side. In truth though, they were both partly right.

In the audience sat a scientist who listened to Shapley and Curtis's arguments with great interest. His name was Edwin Hubble. The former lawyer was soon to deliver the solution to the Great Debate. Curiously enough, the breakthrough came at the same Mount Wilson Observatory where Shapley had done his research.

Thanks to Hubble we have a relatively exact answer to the question of how far a person can see with the naked eye: a bit shy of three million light-years. Our eye covers this distance all the way to an unremarkable spot in the sky, the Andromeda Nebula, M31, the only neighboring galaxy that we can see in the night sky without a telescope. All the other stars that we see are part of the Milky Way. The Andromeda Nebula is also the key to the Shapley-Curtis feud. And more than that—to the entire structure of the universe.

Just three years after the legendary debate, Hubble discovered that the Andromeda Nebula is more than a simple cloud of gas inside of which new stars are born.[20] Within the so-called nebula, he found a star that he was able to use to measure the nebula's distance to Earth. It was the pulsing, periodically blinking light of a Cepheid variable, just like those that Henrietta Swan Leavitt had described. From its light curve, the true brightness of the star and, with that, its distance from the sun could be deduced.

The resultant large distance could mean only one thing: the whole structure had to lie outside the Milky Way. Once Hubble added the results of other observations, he knew that the nebula was in fact a galaxy in its own right. Shapley had been wrong: our Milky Way is only one of many in the universe. Before publishing his findings, Hubble informed Shapley of his work in a letter. It's an open question whether he did this out of maliciousness or a gentlemanly sense of obligation. For his part, Shapley had pointedly criticized Hubble before then and made it clear that he didn't think much of his ideas. But now Shapley admitted his error. When he read the letter and showed it to a student, he told her: "Here is the letter that destroyed my universe."[21]

"The history of astronomy is a history of receding horizons," Hubble wrote in 1936 in his book *The Realm of the Nebulae*.[22] In the case of the Milky Way, however, after the discoveries by Hubble and other scientists in the 1920s, the horizon had yet to open up to the extent that our current knowledge has allowed. The cosmos would grow still further, and where exactly we were located within it was still unclear, because the center was hidden from the gaze of an optical telescope behind all the dust in the disk of the Milky Way.

This changed only in the early 1930s, when the radio telescope opened a new window into the universe for astronomy. In 1932, Karl Guthe Jansky became the first to detect cosmic radio emission when he measured noise signals that were clearly coming from the cosmos, with the strongest signals he received coming from near the constellation of Sagittarius, the archer. Today we know that the Galactic Center, that is, the middle of our Milky Way, lies in this direction.

The Dutch scientist Jan Oort also believed that the center of the Milky Way lay somewhere in this direction. Oort, who lent his name to the Oort cloud that surrounds the solar system and is home to comets, posited a distance of 30,000 light-years. With that he came very close to today's known value of 27,000 light-years. His country-man Henk van de Hulst advanced radio astronomy a step further when he holed up in the observatory in Utrecht during the German occupation of World War II. He predicted that the hydrogen that was so amply prevalent in atomic form in our Milky Way had to emit spectral lines in the radio-frequency range. This should occur, he said, at a frequency of precisely 1.4 GHz, approximately at the range of our cell phone frequencies today.

This was illuminating in the truest sense of the word. Radio waves can penetrate something as thick as a wall, and the dust clouds in the Milky Way don't pose any real obstacle for them either. Now the radio light came shimmering through the dark patches of the Milky Way and van de Hulst and Oort could measure its structure and even discover spiral arms. These are easy enough to see when you're floating above a galaxy, but of course we ourselves are located *in* the galaxy and are looking at it from the side.

In the mid-1950s we finally managed to locate our exact position in the Milky Way. It's a point between the Sagittarius Arm and the Perseus Arm in the so-called Local Arm. From here we travel around the Galactic Center at a speed of 250 kilometers per second. It's a good thing we don't have to adjust our calendars to the Galactic year—it takes 200 million Earth years for our planetary system to make a full turn in our Galactic carousel.

As the planets revolve around the sun, so does the sun revolve around the center of the Milky Way. Today we can track this movement within a few weeks, whenever we look at the black hole in the center with our radio telescopes, as my colleagues Andreas Brunthaler and Marc Reid do regularly.[23] It seems to move across the sky at high speed—a deception, since *we* are the ones who are moving relative to the center of the Milky Way, just like all the stars around us.

In the long term this also has consequences for our view of the sky. In about 100,000 years the seven stars of the famous Big Dipper, part of the much larger constellation Ursa Major, will look different than they do today. The trapezoidal ladle with the handle will look as though someone had smashed it into a wall.

The Milky Way is still a big object of study. The European Space Agency's (ESA) Gaia mission brings in a steady stream of new details about its structure and how it developed. Amina Helmi is a galactic archaeologist and a professor at the University of Groningen. As such, she is the successor to her great predecessors Kapteyn and Oort. In 2018 she brought to light a secret that our Milky Way had kept under wraps since the dawn of time. Remnants of an entire galaxy, Gaia-Enceladus, are still today spinning through our own Galaxy, which swallowed them up ten billion years ago. Capturing this sort of galactic prey led the disk of our Milky Way to grow in size and caused the center to develop a little paunch, the so-called bulge.

But the Milky Way isn't finished developing yet. Many more small galaxies revolve around our Milky Way, and in a few billion years we will fuse together with our equally large neighbor galaxy Andromeda. For our home Galaxy, there are still exciting times ahead.

GALAXIES, QUASARS, AND THE BIG BANG

GALAXIES ON THE RUN

In my intro class each semester we always have a bit of gymnastics. I ask five students to line up, shoulder to shoulder, at a right angle to the wall. The last one places her left hand against the wall with her arm bent and drawn up against her body. All the others place their left hand on their neighbor's shoulder. At my command, I tell them they should all extend their left arms at the same second, thus placing an arm's length of distance between themselves and their neighbor to the left. What happens?

If they all stretch out their arms at the same time, then the student who is standing directly against the wall has to take exactly one step to the side. The student next to her, however, has to take two steps to the side, because he's suddenly two arm lengths farther away from the wall. The student at his side has to take three steps to the side in this same second. And the poor little student at the end? Well, she gets a decent-size shove and goes flying off—five steps in one second are simply too many. Luckily I'm usually able to catch her.

The demonstration is supposed to illustrate what happens when space expands. What happens when a bit of space is inserted between two students—or between two galaxies? They all move away from each other! And the farther away someone is, the faster away he or she moves. This is a simple observation, but when we apply it to space, it changes our conception of the universe every bit as radically as Copernicus, Kepler, or Newton.

Shortly after Einstein published his theory of relativity, he discovered that he had a problem with his universe. The universe wasn't stable. Gravity only attracts. Technically, a universe that is filled with matter should crumple in on itself like a hot-air balloon when the air seeps out. Today we call this the "Big Crunch."

Luckily the equations allowed for a trick: Einstein could insert a free-floating constant into them. It stood for a mysterious force that caused the universe to expand—a kind of antigravity. With this so-called cosmological constant, Einstein managed to save the universe in his model from the Big Crunch—but he wasn't happy about it.

Then things got even worse. In 1922 the Russian physicist Alexander Friedmann wrote to Einstein, saying that he could describe a universe that expanded based on the equations of the theory of relativity—without the mysterious constant. Einstein dismissed this idea. For him the universe had to be eternal and static. At the time there were good reasons for this.

Then of all people, a Catholic priest came along to shake Einstein's fundamental convictions even further. Not only did the priest mathematically describe an expanding universe, but he even claimed that astronomers had already found indications for it.

The priest was Georges Lemaître of Belgium, a former Jesuit school pupil who, after the horror of the First World War, had joined the priesthood and studied mathematics and physics in Leuven. He went on to study under the renowned Arthur Eddington at Cambridge and in Boston at the Massachusetts Institute of Technology (MIT), where he also received his PhD.

Lemaître first noticed the strange characteristics of the galactic nebulae that the American Vesto Slipher had discovered at Lowell Observatory in Arizona. Slipher had measured the speed of the galaxies using the Doppler effect in 1917. We know the effect from the realm of acoustics: If an ambulance drives past us with its siren blaring, we hear a higher-pitched sound for as long as the vehicle is driving toward us. Once it has passed and is moving away from us, we hear a deeper-pitched sound. What is true for sound is true for light as well. If galaxies are flying toward us, the light is compressed and more blue; if they are flying away from us, it is stretched out and more red. Naturally it always flies at the speed of light in both directions, but our perception of the color changes. So, if you measure the fingerprint of the atoms in the light of the galaxies with a spectrograph, then you can also measure the slightest color shifts and thus the speed of the galaxies in the direction you're looking from.

The result: with the exception of our neighbor, the Andromeda Galaxy, the light was mostly redshifted. Almost all the other galaxies were moving away from us! This was more than strange, and couldn't be coincidence. Imagine the confusion in a large ballroom. The hall is filled with couples gliding around the dance floor. Shouldn't the number of couples moving toward you be about the same as the number moving away from you? What if it turns out that everyone is moving away from you? Are you that unpopular?

Lemaître's answer: it's not us; rather, the entire universe is expanding, and the light along with it. By correlating the galaxy speeds that Slipher had calculated with the distances that Hubble had calculated, Lemaître found that the galaxies fly away from us at greater speed the farther away from us they are. The galaxies that are farthest away move fastest—just like the poor little student at the end of the row in my lecture.

The sense of relief was immense. This hectic flight of the galaxies isn't the result of some unpleasant repellant quality of our own Milky Way; rather, other observers in other galaxies would see the same thing we were seeing. Unlike the wall in my classroom, the Milky Way isn't anchored down anywhere in space, nor is it resting at the center of the universe; rather, it's moving somewhere in the crowd of the cosmic ballroom along with all the others. The whole cosmic dance floor, ballroom included, keeps expanding and expanding.

You could also imagine it like this: If the dance floor were the outside of a giant balloon, the dancers would be dancing on its surface. If the balloon inflates, there's more and more space, and all the dancers move away from each other. Only the individuals locked in each other's arms remain together, like the Milky Way and Andromeda. Their mutual attraction is stronger than the force of the expanding universe.

Lemaître published his results in 1927 in French, citing Hubble's distance measurement data. Two years later Hubble published the same correlation using almost the same data, only he published his work in English. But he left out Slipher, whose measurements he used, and made no mention of Lemaître either, whom he had personally spoken to. Scientific historians and contemporaries of Hubble's both say that he "was very selective in referencing, failing to mention in his publications those of his colleagues."[1] This is putting it kindly. In science, the mentions and acknowledgments you receive from your colleagues are the one hard currency. Behavior like Hubble's is unfortunately not uncommon, but it is highly unethical.

Science is sometimes like Homer's ancient heroic epic *The Iliad*: more important than your deeds and even your own life are the stories that are told about you afterward. Hubble was concerned with reserving a special place for himself in history, and he succeeded. The famous space telescope is named after him and the law of expanding space was for a long time simply called Hubble's Law. Only in 2019 did the International Astronomical Union vote to rename it the Hubble-Lemaître Law.

The Hubble-Lemaître Law was crucial in expanding the horizons of the universe. With it, it was now possible to measure the distances between Earth and the most remote galaxies. Billions of light-years no longer posed a problem. So long as you could find the spectroscopic fingerprint of the atoms emitting light in a galaxy, then the spectral line redshift was a measure for its distance.

Albert Einstein didn't at all agree with this new development. Looking backward, this expansion would mean that long ago the whole universe would have to have been pressed together into a single point. Once again his equations had resulted in a singularity in time and space—same as with black holes. This meant that the universe had to have had a beginning! Lemaître was the first one who dared to give voice to this thought, and spoke of a primeval atom from which the young universe was born billions of years ago, as if from an egg.

Einstein didn't like the taste of that either. Didn't that sound suspiciously like the wishful thinking of a priest? Didn't that idea come from Biblical notions about creation? The Catholic Lemaître stood under general suspicion. Scientists remained skeptical, and some even scoffed at his model, calling it the "Big Bang." Indeed, the term was originally used in a negative sense, but because ultimately the thought behind it was a valid one, it remained in use. In German, *Urknall,* meaning "original" or "primordial" bang, has entered the vernacular—a more fitting term, I think.

In a long conversation Lemaître tried to convince Einstein that his static universe didn't work. But it took a long time for the Big Bang theory to be completely accepted. As a young scientist, I still met old eminences who firmly rejected this idea. "The Big Bang lets the creator jump right back out of the coffin," or so they feared. But here we had history repeating itself with the roles reversed. While in Copernicus and Galilei's day it was the Vatican that rejected the new model of the universe, in Lemaître's it was Pope Pius XII who in 1951 became one of the first to support the new theory of the expanding cosmos.

They say an old theory dies with its last critics, and indeed that's how it was in this case as well. Today the model of the dynamic, expanding universe is completely accepted among scientists, even if the mystery of the Big Bang still has yet to be uncovered.

A NEW LIGHT: RADIO ASTRONOMY

For thousands of years, humans could only look at the heavens with the naked eye. Then, starting in the seventeenth century, we were helped by optical telescopes. Ninety years ago, however, a completely new technique was adopted. In that brief span of time it was to revolutionize the study of space. When Karl Guthe Jansky discovered cosmic radio signals in 1932, we instantly saw the entire universe in a completely different light—literally, because for the first time we weren't using visible light, but rather light from a different range of the electromagnetic spectrum. For astronomers this meant stepping into completely new territory, which took some getting used to at first; several turned up their noses. It took some time for this new discipline, radio *astronomy*, to become established within the broader discipline of astronomy, and for its instruments to be accepted as radio *telescopes*. The image-producing components of optical telescopes are generally made of different layers of glass; radio telescopes are made out of steel.

Today we search the heavens for the full electromagnetic spectrum, and for that purpose we employ radio, infrared, optical, x-ray, and gamma-ray telescopes. We receive radio waves with a frequency of 0.01 GHz that are as big as a house. Or gamma rays with frequencies of 100 billion GHz that are 100 million times smaller than an atom. One gigahertz equals one billion oscillations per second; it is the type of radiation that we use for Wi-Fi. Visible light oscillates at 500,000 GHz. One could compare the radiation emitted by the universe to a cosmic symphony, where every individual frequency corresponds to a note on light's musical scale. The instruments we have today extend over a frequency range of sixty-three octaves, which would equal a piano with a keyboard almost 12 meters wide. Until the beginning of radio astronomy, however, we only heard the light-music of the universe in a single octave. With radio telescopes, the lower notes were added little by little and gave the universe a completely new sound. Suddenly the sky, lit up by radio-frequency radiation, shone not only with stars but also with black holes and the light of the Big Bang. Later, x-ray and gamma-ray telescopes brought us the higher notes as well.

The breakthrough for this new astronomy came after World War II. The timing wasn't coincidental: air warfare had spurred on the development of radar. Of all things, the slaughter of the war had produced the necessary technology. For all the fascination that radio astronomy inspires, we must never forget its bitter origins. After the war, radio antennas, receiving dishes, and transmitters were available in great numbers, and among astronomers there ensued a race to claim them.

In the years that followed, research was dominated by giant radio antennas that engineers had once built for radar facilities. In England a group of former Royal Air Force soldiers under the leadership of Bernard Lovell began building a giant telescope with a diameter of 76 meters at Jodrell Bank. As a result of a calculation error, the dimensions were completely ill-suited for its actual purpose. The project lapsed into financial difficulties, and Lovell feared he would go to prison. But the Sputnik shock of 1957 saved the telescope, because the group was the only one in all of England that was able to receive and interpret radio signals from the first Russian satellite. Of course, it wasn't with the dish that they managed to do this, but rather with a simple wire.

The Dutch likewise began exploring the sky in this new range of light. They first started with a German radar facility, then built a 25-meter telescope in Dwingeloo, which was used to measure the 21-centimeter-wavelength radiation produced by hydrogen that Henk van der Hulst had predicted.

A radio dish 64 meters in diameter built in Australia near the small town of Parkes in New South Wales made history with its scientists' dramatic effort to be the first to pick up television footage of the Apollo 11 moon landing.

In the '70s, German radio astronomers built the largest movable radio telescope in the world, with a diameter of 100 meters, in the sleepy town of Effelsberg near Bonn, then the capital of West Germany. As a PhD student at the Max Planck Institute for Radio Astronomy, which operates this instrument, I used it to make my first radio-astronomical observations.

There was only one radio telescope that was larger, and that was the 300-meter dish at the Arecibo Observatory in Puerto Rico, which had been built by the US Department of Defense in the '60s and later handed over to astronomers. It was built in a natural basin and was completely immovable. As a result, you could only observe a small part of the sky. The facility became famous thanks to the James Bond movie *GoldenEye*, in which the villain fills it with water. In 2020, cables snapped and wrecked it, putting it up for demolition.

Around the same time, the Americans were building a movable radio dish 90 meters in diameter. It was located in the town of Green Bank, in a very rural part of West Virginia that was declared a radio-quiet zone. Today the town is very popular among people who are afraid of radiation. In the '90s the telescope collapsed overnight due to metal fatigue. The day before, a colleague of mine from Bonn[2] had snapped the last photo ever to be taken of the telescope, and the next morning he took a photo of the rubble heap. As a rule we radio astronomers aren't superstitious, but in the years that followed everyone always got a bit nervous whenever this colleague took out his camera.

The Green Bank Telescope was rebuilt, and this time its diameter was effectively one meter longer than that of the Effelsberg 100-meter Radio Telescope in Germany. I've never quite been able to grasp the scientific justification for the extra meter, but it was clear that this technology had reached its limit. No one could or would build still larger telescopes.

Nevertheless, we astronomers urgently needed larger facilities to produce clearer images. The image resolution of a telescope depends on the wavelength of the light and the telescope's diameter: the larger the telescope, the clearer its gaze. But the image also gets more and more blurry the greater the wavelength at which the observations are conducted. Radio astronomy works with much greater wavelengths than optical astronomy, which means the 100-meter telescope in Effelsberg doesn't see any more clearly than the human eye. You can't spot a black hole with that. If you want clear images, you have to think bigger. The solution came in the form of radio interferometry. This is a technique for linking several telescopes together in order to produce the equivalent of one giant telescope.

Ruby Payne-Scott of Australia pulled off the first successful radio-interferometric measurements after the Second World War. All she had was a single antenna, but she used the ocean's surface as an additional radio reflector. In 1964, Martin Ryle built the One-Mile Telescope in England and later received the Nobel Prize in physics for successfully linking together three radio dishes to form a single large telescope. Other radio astronomers kept refining this principle to produce ever clearer images. In the Netherlands, a network of fourteen 25-meter dishes was built—on the site of the former Westerbork concentration camp, of all places. And in New Mexico in the United States, the Very Large Array (VLA) came into being, consisting of a total of twenty-seven parabolic dishes that could be arranged in various configurations over thirty-six kilometers. Each individual dish was twenty-five meters in diameter, which meant that VLA scientists were ultimately able to use a telescope that was effectively larger than the entire Boston metropolitan area. For many decades it was one of the most productive instruments in the entire field of astronomy.

Finally, we started linking together radio telescopes all over the world—the idea was to build facilities that were as big as the globe and could produce the clearest astronomical images possible. The method was given the clunky English name Very Long Baseline Interferometry, which astronomers tend to abbreviate and refer to as VLBI. The very long baseline results from the telescopes being positioned very far apart from one another. With this technology, we now had global telescopes. In the end, it was this very technology that enabled us to capture the image of the black hole.

QUASARS: EN ROUTE TO THE MONSTERS OF MASS

Thanks to radio astronomy, astronomers were able to make completely new discoveries. It was as if in addition to touch, smell, taste, sight, and hearing they now had a sixth sense. Soon they began to search the sky systematically for radio sources. All of a sudden astronomers were finding thousands of new celestial objects, and no one knew exactly what they were. It was assumed at first that they had to be stars. What else could they be?

In Australia, John Bolton picked up a radio source from the direction of the Messier object M87 and claimed it must be part of our Milky Way, even though he was silently convinced that M87 was its own galaxy! For fear of being ostracized,[3] he didn't dare inform his colleagues that this radiation was reaching us from many millions of light-years away—because if an object was that far away, and we could still detect it, how strong must the radio luminosity be? What celestial body, what galaxy, what mysterious object in space could produce such a high amount of radiation? The thought was too radical.

Just ten years later Bolton's fear was gone, and the existence of so-called radio galaxies had long been accepted. Among them were M87 and the galaxy Cygnus A, which, if the Hubble-Lemaître Law was to be trusted, had to be a full 750 million light-years away from Earth. Great excitement spread among astronomers, because this radio light that we had been measuring for just a few years now allowed humanity to look into the deepest depths of space and, by extension, deep into the past of the universe.

Researchers at Cambridge produced a large catalog of every radio source. The first version of the catalog was too small, the second version contained a lot of mistakes, but the third version, called 3C, served as the foundation for many studies. New radio stars and radio galaxies were simply numbered serially. But no one had even a vague idea of what exactly was emitting the radiation. The images of these mysterious objects in the sky were still extremely blurry, the determination of their position extremely imprecise. It was ascertained that the radiation itself was produced by electrons moving close to the speed of light, which were diverted within cosmic magnetic fields. Astronomers knew the process from particle accelerators on Earth, which are called synchrotrons, and thus this radiation was called synchrotron radiation.

Certain sources were stretched out lengthwise and looked like barbells, others appeared to be small and point-shaped, like a star—and indeed, switching to a different wave band led to the discovery of something in the visible light range at the position of object 3C 48 that did in fact resemble a star. But spectrographic analyses of this apparent star produced more questions than answers: The object 3C 48 showed emission lines of unknown wavelengths. The bar code in light couldn't be matched with any known element. Had a new element been discovered in space?

John Bolton and his coauthor, Jesse Greenstein, briefly wondered whether it might possibly be the redshifted light of hydrogen, but that also seemed too extreme. Because if it was, then this object would have to be at a position in space that was approximately 4.5 billion light-years away. "I had a reputation for being a radical and was afraid to go out on a limb with such an extreme idea," Greenstein said later.

The strongest argument against the unbelievably-great-distance hypothesis was that the luminosity of this light source would change dramatically within just a few months. This couldn't be a galaxy! How could billions of stars located hundreds of thousands of light-years apart decide to vary their pulsation period all at once so that their collective light would grow brighter and darker almost simultaneously within the span of a month?

Imagine if all eight billion people in the world were to clap their hands together at the same time. I wouldn't hear a single, brief burst of sound, but rather a quiet, long-lasting rumble, because of course the sound would be spreading out from points all over the Earth and would never reach a single listener all at once.

On the other hand, I can at least estimate the size of the sound source from the duration of the sound and the speed of sound. The shorter the duration of the sound, the smaller the space it must be coming from. If I hear a clapping sound that lasts for a second, then all the people must be sitting in a stadium, because a stadium is just about the size of a sound-second, that is, the time it takes sound to travel one second. It could also be someplace smaller, too, of course. It's the same with variable light sources: if the variation occurs within a month, the source can't be larger than a light-month. This is much smaller than the distance between us and the closest star. So that meant 3C 48 had to be a star then, right?

After that, astronomers turned to the next-brightest radio source in the catalog: 3C 273. In order to determine its exact position, radio astronomers at Parkes Observatory in Australia used a trick: they got the moon to lend a hand. By chance its orbit happened to pass in front of the quasar's position in the sky. When the moon moved in front of the radio source, the signal briefly vanished from the reach of the large antenna. It was like a solar eclipse, only here it wasn't the sun that was being covered by the moon, but rather the mysterious radio object.

At the exact moment of the radio signal's disappearance the astronomers measured their first coordinate of the object: it had to lie somewhere along the near edge of the moon. They got the second when the far edge of the moon cleared the object and let the radio signal from 3C 273 come through again. Since we know how large the moon's diameter is and where exactly it is located, it was possible to calculate the exact position using the point of intersection between the two coordinates.

Incidentally, 3C 273 might be one of the brightest radio sources in the sky, but in terms of cell phone frequencies it's only five times brighter than an LTE phone on the moon that we measure from

Earth. Once the position of 3C 273 was known, Maarten Schmidt, a Dutch astronomer who worked at Caltech in Pasadena, started investigating this region with the Mount Palomar telescope. He found a fairly bright star, so bright that today even an amateur astronomer with a decent telescope would be able to track it down in the Virgo constellation. Schmidt immediately analyzed the spectrum of light emitted. Again the bar code was highly strange. After six weeks he finally recognized a pattern and was certain: This was the hydrogen spectrum of an object that had to be a scarcely imaginable two billion light-years away. The expansion of the universe had stretched the light out so much that it was redshifted by 16 percent and appeared in a spot where no one had expected it to be.

Schmidt's data were so good that he was emboldened to publish them. He might not have known just what this cosmic star was supposed to be, but that didn't scare him off. Because the object only looked like a star but likely wasn't one, he simply called it, for lack of a better term, a "quasi-stellar radio source," or QSR. In astronomers' slang that became "quasar." "It was as if suddenly the blindfold had been lifted and we realized that a star isn't a star," Schmidt said later.[4]

Today it's hard to imagine the excitement that this discovery caused. The horizon of the visible universe had expanded immensely; outer space had literally exploded.

The whole universe seemed to change and develop with time. Ten billion years ago it was the age of the quasars—back then their activity reached its peak. Their number had increased rapidly in the first four billion years of our universe and illuminated all of space. Then, in subsequent eras of the universe, the quasars burned out one after the other.

But what exactly was 3C 273? The conclusions drawn from the observations were dramatic. If 3C 273 could still be seen so brightly from Earth at this massive distance, then it had to shine a hundred times brighter than a whole galaxy. And if this quasi-star flickered within a period of a few weeks and months, it couldn't be much bigger than a light-month—probably only as big as a single solar system.

So it began to dawn on the astronomers that 3C 273 must be a very uncanny place indeed. This object gave off an unimaginable amount of energy, and all this energy was originating from a relatively small point in the universe. How can you create so much energy in such a tiny region of space? Whatever this quasar might be, it was driving even the cleverest astronomers close to the point of despair. No one had ever encountered such gigantism in astrophysics before.

Some scientists' thoughts quickly turned to the greatest of all forces in space, gravity. For something to shine that brightly, its mass had to be unimaginably huge. Sir Arthur Eddington had originally developed this argument for stars. Light, too, exerts pressure. If a star were to shine too brightly, it would burst—just like a balloon pops if you inflate it too much. Given its luminosity, only a massive amount of gravitational force could hold such a gigantic celestial object together.

If you used the Eddington argument to calculate the minimum mass necessary to hold a quasar together, the result was a value of almost a billion suns. It was enough to make you crazy: The light of a billion suns and the mass of a billion suns was supposed to be able to fit inside a single solar system?

Six years after the discovery of quasars, the English astrophysicist Donald Lynden-Bell was trying to figure out how to resolve these contradictions. What if, in the center of galaxies, there were supermassive black holes? Not a small stellar black hole born of a single supernova, but rather billions of dead stars melted together to form one giant monster. Only such an object could put out so much energy without simultaneously being torn apart. And it would be small enough, too. After all, Roger Penrose, a British mathematician and theoretical physicist, had just shown that black holes could naturally form within the theory of General Relativity.

But how can a black hole emit light? Isn't it supposed to be black?

Yes, the black hole itself is dark, but not the gas that is attracted by it and about to disappear into it. It is in fact speeding with unbelievable energy toward the black hole and is heated up by gravitational energy, angular momentum, and magnetic friction. Plus, black holes are unbelievably efficient; they cause almost everything around them to move at nearly the speed of light.

Let's imagine a fist-size bocce ball made of metal. If we throw it onto the bocce court, it hits the ground with a nice thud and leaves behind a small depression. If we place this same ball in a cannon and shoot it off at a speed of one kilometer per second, the ball can break through walls. Now, what happens if we let it fall toward a black hole and it reaches almost the speed of light? It would be 300,000 times faster than the ball shot from the cannon. But because the kinetic energy grows in square proportion to the velocity, the ball would now have one hundred billion times more energy. The total energy of the bocce ball would thus equal about ten billion kilowatt hours. With the energy from the impact of a single one of these balls you could supply the homes of three million German families with electricity for a year.

This sounds unimaginable—but black holes are capable of such things. If dust and gas enter the gravitational field of a black hole, a turbulent disk of gas and magnetic fields is formed, similar as with new stars—the so-called accretion disk. And along its inside track this giant whirlpool spins around the black hole at just shy of the speed of light. The gas heats up as a result of the magnetic friction and emits a blazing light. The so-called black hole shines like a bright-blue star. A small portion of the hot plasma flowing into it is shot out into space by the magnetic fields in a giant glowing jet. In appearance these jets do in fact resemble the exhaust trail of a jet plane. As a result only a few lucky particles manage what is denied all the others—they're only just able to escape the black hole. As in the sun's corona, the particles are accelerated in the magnetic fields and give off bright synchrotron radiation. It is precisely these hot, radiation-emitting, magnetized, and focused jets that we see escaping the quasars in our radio telescopes.

The efficiency of these gravitational maelstroms and their jets is enormous—it's up to fifty times greater than the nuclear fusion that occurs in stars. Black holes are thus the most efficient power plants in the universe. Instead of dropping my bocce ball I could pour a liter of water into a black hole and produce enough energy for a city of millions for a year. The water I've got, but unfortunately I lack a black hole within easy reach. Otherwise all our energy problems would be solved just like that.

Quasars' thirst is gigantic: they devour forty-five times the total amount of water on Earth every second—this equals the mass of an entire sun per year. Black holes don't operate in a particularly sustainable fashion either—the water swallowed up by a black hole can't be recycled. What's gone is gone. A black hole is extremely ego-tistical. With every sip, it only gets heavier, bigger, more attractive, more dangerous.

With 3C 273, astronomers indirectly discovered the first black hole. But not everyone in the scientific community shared the belief that black holes existed, not by a long shot. Decades would pass before this theory became paradigmatic. Some believed quasars were stellar objects that were spat out of galaxies. For today's astronomers such theories are bizarre, but they were actually discussed, because the path to ultimate proof was still a long one.

MEASURING THE BIG BANG

Concurrent with the discovery of quasars, our understanding of the cosmos as a whole also began to develop rapidly. In 1964, Arno

Penzias and Robert Woodrow Wilson at Bell Laboratories began using a telecommunications antenna to listen in on the radio static of the heavens. The antenna resembled a supersize ear trumpet. At first they didn't at all like what they were picking up. From every direction they received a weak, persistent, bothersome static. They checked all the cables, chased the pigeons away, and cleaned their droppings off the antenna—but the signal was still detectable. In the end they wound up inferring the existence of a cosmic microwave background, because the radiation came steadily from space. Its quality was exactly like the thermal radiation that would come off a black, opaque fabric that spanned the entire sky and registered at about 3 degrees on the Kelvin temperature scale. (This equals −450 degrees Fahrenheit, only 9 degrees above absolute zero, the point where nothing moves at all anymore!) For that reason it's also known as 3K or 3-degree radiation. This was what was left over from the fireball that was the Big Bang, and for discovering it Penzias and Wilson would later receive the Nobel Prize in physics.

In the early phase of the universe, space was filled with extremely hot and opaque gas. Protons and electrons were flying wildly this way and that. But the more the universe expanded, the cooler it got. Three hundred eighty thousand years after the Big Bang, the universe was still about 3,000 degrees Kelvin—as hot as molten steel, but just cool enough that protons could capture electrons with the force of their electric charge and build the first atoms. The universe became a sea full of hydrogen gas that was now transparent.

The free-floating electrons, which before then, like tiny antennas, had absorbed all the light, were suddenly trapped in the atoms; the fabric was pulled away, the light was released, and ever since it has been shining unencumbered all the way to us. As a consequence of the universe's expansion, we have gotten farther and farther away from a portion of this light. Light waves that still reach us today have, over the course of their 13.8-billion-year marathon through the expanding cosmos, been stretched out by a factor of a thousand and have since cooled down. Instead of waves corresponding to a temperature of 3000 Kelvin, today we only receive the supercold 3 K radiation. What reaches us is a chilly wisp of the Big Bang's original thermal radiation. But through it we look back into the dawn of cosmic time, when the universe resembled an impermeable, hotter-than-molten-steel furnace. This is as far back and as deep as we can look, no farther. This discovery of the cosmic microwave background, which came as a surprise to many, became the decisive proof for the Big Bang model: we're looking the beginning of space and time in the face.

In the '90s the COBE satellite took extremely precise measurements of the cosmic radiation and found the slightest variations in luminosity. These stem from waves in this primordial hydrogen ocean and were the predecessors of the first superclumps that over the course of the extended history of the universe drew together into galaxy clusters and galaxies. Thanks to the Wilkinson Microwave Anisotropy Probe, WMAP, and the Planck satellite launched by NASA and the ESA, respectively, along with many other experiments, these seeds of today's galaxies have since been measured in detail and provide us with minute insights into the history and structure of the universe.

Astronomers involved in the large-scale surveying of the sky that began in the late 1980s have in fact discovered that galaxies across the universe do not appear evenly spaced but rather are spread out in a filigreed pattern or gathered together in large agglomerations. It turns out that galaxies are more social and attractive than you might think, and they're often stacked one on top of the other in large clusters.

Naturally the individual galaxies in these galaxy clusters don't stand still. Rather, they move around and mingle among one another under the influence of gravity. Often they're flying past one another at over 1,000 kilometers per second. Seen over billions of years, galaxies might even be said to move like a spritely school of fish: sometimes they'll even slip into two or three other galaxies to build a new, larger galaxy that takes the form of a massive ball or a thick cigar. We call them elliptical galaxies; the M87 Galaxy is one of them. Their stars never collide, though, or as good as never, because they're all located well apart from one another. They only feel the influence of each other's gravity.

The heavy galaxies sink into the center of the cluster and increase in size. Even their black holes melt together—and so the largest and heaviest galaxies often sit in the center of galaxy clusters and harbor the largest black holes in the universe. They're the monsters among the giants. Our neighbor galaxy M87 was also formed in this manner. Among all the galaxies and black holes that make up the superheavyweight class in the universe, M87 is the closest to us.

Really, though, these galaxies were moving way too fast—so it appeared to Fritz Zwicky, a Swiss astronomer conducting research at Caltech in Pasadena way back in 1933. The gravity of the stars wasn't great enough to hold the careening galaxies in place—really they should be scattering off in all directions. But that's precisely what they weren't doing, which meant some mysterious force must hold them in place. If it was gravity, then there must be some mysterious dark matter there that you couldn't see, and moreover there must be five to ten times more of it than the normal matter known to us.

In the 1970s the astronomer Vera Rubin measured how fast galaxies move using optical telescopes and the Doppler effect. They seemed to be rotating a bit faster than they had a right to. This was confirmed by the Dutch scientist Albert Bosma, who investigated this phenomenon with the new radio interferometer in Westerbork. He saw gas from which no stars had yet formed and that extended much farther than the galaxy that one could see with optical tele-scopes. Here too everything was rotating far too fast. Galaxies had to be full of this dark matter that held them together. Without it, galaxies would fly off like soup bowls on the lazy Susan at a Chinese restaurant when you spin it too quickly.

To this day we don't know what dark matter is. Some astronomers think the theory is nonsense and claim that dark matter doesn't exist. Rather, they say, the laws of gravity are simply wrong when applied at a galactic scale. These objections notwithstanding, most astronomers today assume that dark matter consists of a still-unknown family of elementary particles.

Things got even more confusing in the 1990s when the sky was systematically searched for supernovae, whose brightness one could measure quite well. Now it turned out that they shone a little bit less brightly than one would expect given the expansion of the universe and the Hubble-Lemaître Law. Were they farther away? If so, the universe would have to have expanded more quickly than previously thought. Since then, dark energy has also been a part of our physical-astronomical understanding of the world: an unknown, mysterious energy that causes the universe to expand ever more rapidly. This dark force was already hidden in Einstein's equations as the cosmological constant, but at some point he had dismissed it, calling it his "biggest blunder."

The most up-to-date simulations and measurements of the cosmos show that, of the total matter in the universe, about 85 percent of it belongs to the dark side. Fifteen percent is the normal, so-called baryonic matter that we're familiar with. On top of that, measured across the entire universe, today dark energy comprises twice the amount of energy that is contained in dark and normal matter combined. Mass is, after all, the equivalent of energy, according to Einstein's famous formula $E=mc^2$. Altogether then, only about 5 percent of the total energy in the universe is contained in the form of matter as we know it on Earth—the atoms and elements of the periodic table. As to the origin of the massive remains, we're literally in the dark.

Astronomers often describe this discovery as another Copernican revolution: humans are neither at the center of the universe, nor of the Milky Way, nor of our solar system; what's more, our bodies and our entire world are made of a form of matter which, in the context of the entire universe, can be considered exotic. I like to look at it the other way, however: we've learned that we're woven of a very special cloth.

There's no immediate connection between dark matter and dark energy on the one hand and black holes on the other, though they might seem equally mysterious and dark. Dark matter can certainly fall into black holes and cause them to grow. This probably only happens in very small measures, though, because the dark matter in the center of a galaxy is very thin and spread out. Dark energy as well can only be read on the large scale of the universe, and in theory shouldn't change the structure of black holes—just as a breath of air can't really topple Mount Everest in the short term, even though the entire mass of air on Earth is ten thousand times heavier than this one mountain. All the same, the unknown nature of dark matter and dark energy calls our attention to gaps in our understanding of physics. A new theory of space and time that took into account dark matter and energy could also change the equations that govern black holes.

PART III

The Journey to the Image

My personal journey to the
Event Horizon Telescope and the
first image of a black hole

THE GALACTIC CENTER

THE FASCINATING GARBAGE CHUTE

I grew up near the Südstadt neighborhood of Cologne, today filled with students and a ten-minute walk from the University of Cologne's physics institute. Later I would attend my first lectures there, and held a temporary lectureship there. But when I was little, my world was the sidewalk in front of our building, where a group of kids was always playing. The street was still paved with cobblestones, and the highlight of the week was when the garbagemen in their orange work uniforms would skillfully roll the big trash containers from the back courtyard through the center passageway to the big garbage truck out front. More than anything I wanted to be a garbageman and to get to drive a truck like that, which grabbed the giant containers and gobbled up the garbage. That the men were able to operate such a powerful machine simply by raising a lever was fascinating to me. My career choice was clear: anything to do with great big machines!

Later, though, I ended up going with physics, and in my final thesis for my master's degree I set out to tackle black holes. There turned out to be surprising parallels to my childhood fascination. Black holes are essentially cosmic garbage chutes, and they exert an unbelievable force of attraction—not just on large stars, but also on little college students. I wrote my master's thesis under Professor Peter Biermann. In his dealings with students he made a point of being extraordinarily generous. He was always getting crazy ideas and loved to discuss them with us. Biermann knew the whole world; he traveled a lot and knew what subjects were in vogue in astronomy. Even more importantly, while he was off traveling all the time, we were able to work undisturbed! My own doctoral students will be thoroughly familiar with this arrangement—I'm on the go a lot, too. Nevertheless, Biermann had remained a physicist of the old school who was quick to grab a piece of chalk and write up all the important calculations and estimations on the blackboard and who could work out logarithms in his head. His father, Ludwig Biermann, had been director of the Max Planck Institute for Physics and Astrophysics in Munich and had published important work on the sun's magnetic field. The Biermann home hosted science luminaries like Werner Heisenberg and Otto Hahn, whom the young Biermann knew simply as "Uncle Otto."

The first time I was seized with fascination for black holes wasn't in the classroom, however, but rather when I read an article by Charlie Townes and Reinhard Genzel in *Spektrum der Wissenschaft (Scientific American in Germany)*. In this article the authors speculated that one of these remarkable supermassive black holes with a mass of approximately two million suns could even be lurking at the center of our Milky Way.[1]

I was immediately captivated. The article made it clear to me that a lot of exciting things were happening in the field of astronomy. I thought particle physics was fascinating, too, but at the time it wasn't advancing with any kind of speed. It was all about building large particle accelerators, but decades would pass by the time they were finished and producing results. A black hole at the center of our own Galaxy—this somewhat mysterious idea appealed to me right away.

A further incentive was that gravity was the last force that was still not understood. It stubbornly resisted every attempt to unify it with quantum physics and other natural forces. Gravity is the big stumbling block on the road to a unified theory. True, I had no idea what such a theory might look like, but it couldn't hurt to be on the lookout for one. Maybe I could add a little stone to the edifice of physics. When you're planning to build a house, it's a big help to know what kind of house you'd like to build. Planning a career is no different: you have to know where you want to go. I thought to myself that if there was any place where something exciting was happening, then it was definitely on the edge of black holes.

Those were adventurous times. I had moved on to the Max Planck Institute for Radio Astronomy in Bonn by then and was sharing a tiny one-person office with two other colleagues. One of the desks even jutted out into the hallway. For my final thesis, Peter Biermann had posed me the theoretical question of whether the accretion disk of a quasar might also send the same kind of wind out into space that we know stars to emit. Astoundingly enough, rotating disks of matter around supermassive black holes share a lot of similarities with hot, flattened stars. As such, the high radiation pressure of the light should blow away the outer layers of the disk. We've seen these extreme winds coming from very hot stars, which fling a lot of matter out into space. With black holes you also have to take into account that light, too, is diverted and focused by the warping of space. And so I calculated how gas moves in the light of quasars and how the black holes at their center divert the light.

The topic was exceedingly interesting to me, and the effect was in fact discovered later on, but at this point in time it was a purely theoretical exercise. In 1992 I began my doctoral thesis on the same topic. Peter Strittmatter, director of the Steward Observatory in Arizona and a close colleague of my thesis advisor, came to visit in order to plan a new submillimeter radio telescope in Arizona with the Bonn crowd. I proudly told him about my project. He listened politely—while trying not to yawn. It wasn't just him. My topic didn't really seem to have grabbed anyone else's attention either.

Nevertheless, 1992 was an exciting year, a year that would set my life on a new course. Our daughter glimpsed the light of this world for the first time, and at the center of the Milky Way as well, a new world suddenly came into view.

THE DARK HEART OF THE MILKY WAY

No sooner had quasars been discovered and the idea of black holes developed than people started to think further. If, billions of light-years away, in the early turbulent years of the cosmos, giant black holes had been present at the centers of galaxies, then they couldn't simply have disappeared in the time that had passed since then, right? And if there are a few galaxies with black holes, why shouldn't every galaxy have a black hole?

Soon astronomers were starting to notice some curious goings-on at the centers of galaxies quite near us at a distance of only 50 million light-years. The cores of these galaxies seemed to shine brightly and to spit out radio plasma. Hot, luminous gas was spinning around their centers. These remarkable galaxies had been known since the '40s and were named Seyfert galaxies after their discoverer, Carl Seyfert. Were black holes working their mischief here as well? In the

'70s and '80s, astronomers had put together a whole menagerie of galaxies suspected of having black holes. Together they were called active galactic nuclei (AGN), and this whole area of research was buzzing with activity. Was there one of these mass monsters at the center of our own Milky Way too?

The British astrophysicists Donald Lynden-Bell and Martin Rees had voiced this very suspicion in 1971 and predicted that a continental radio interferometer—that is, a VLBI experiment—could discover a compact radio source like a black hole at the Galactic Center.

Radio astronomers promptly began the search, and just three years later Bruce Balick and Robert Brown did in fact discover such an object at the center of our Milky Way using the radio interferometer in Green Bank, West Virginia. In so doing they beat out Ron Ekers and Miller Goss's team at the University of Groningen by just a hair. Ekers and Goss had combined data from an interferometer in Owens Valley, California, with data from the brand-new radio interferometer at the Westerbork former concentration camp site in the Netherlands and confirmed the discovery of the mysterious radio object.

This new radio source is located in the middle of a region that is known as Sagittarius A and is seen as the heart of the Milky Way—the Galactic Center. Sagittarius A is far and away the brightest radio source in the Sagittarius constellation. The second-brightest radio source is called Sagittarius B. Even several years after Balick and Brown's discovery, people were still writing articles about the "compact radio source in the Galactic Center." Finally it got to be too much for Robert Brown and he came up with the abbreviation Sagittarius A*. The asterisk was meant to convey only that it was a very exciting object. Because astronomers are lazy and don't like typing, they generally abbreviate it even further and speak of Sgr A* (Sag A*).[2] Naming practices in astronomy might make science reporters tear their hair out, but we astronomers find them completely natural and totally normal.

Soon scientists were taking VLBI measurements of Sgr A* and hoping for clearer images, but they came away disappointed. The object looked utterly boring: an almost round, slightly flattened splotch. People hadn't pictured a black hole as being so unspectacular. In the following years ever higher frequencies were used that promised considerably clearer images, but still all radio astronomers could see was a splotch—albeit a smaller one. Then it dawned on them: the Milky Way acted on radio-frequency radiation like a giant pane of frosted glass, making fine structures appear blurry. They could only see an unfocused image of what was going on at the center of the Milky Way—hot gas and dust in the galactic disk blocked our gaze. How disappointing!

This was even more true of visible light. The thick gas and dust clouds in the disk of our Milky Way didn't just scatter the visible light, as they did light in the radio-frequency range; they absorbed it completely and blocked off every glimpse behind this "curtain." Would the Milky Way keep its secret forever?

I had just begun my PhD dissertation when suddenly this veil was lifted. Experts from all over Germany came together at a spontaneous miniworkshop in Bonn and reported on their most recent, still-unpublished findings about our Galactic Center. I was electrified.

Working under the former director at Bonn Peter Mezger and his colleague Robert Zylka, the group took measurements of Sgr A* at a wavelength of 1.3 millimeters for the first time in 1988. This was the exact same millimeter wave band that we would later use for our image. True, they had only a single telescope and couldn't produce any clear images, but it seemed that Sgr A* shone surprisingly bright in this range. What was even more astonishing, though, was that at higher frequencies, in the far-infrared range, the radiation cut off dramatically and was no longer detectable. So what was creating this millimeter-wave radiation? Ultra-hot gas in the vicinity of the black hole or just a warm dust cloud farther away?

In the '90s, collaborators at the Max Planck Institute in Bonn and the Haystack Observatory at MIT did pioneering work in the development of VLBI in the millimeter wave band. My colleague in Bonn Thomas Krichbaum had just taken the first VLBI measurements of Sagittarius A* at 43 GHz, or a wavelength of seven millimeters, and here at the workshop he discussed his brand-new results. These were the clearest images at the shortest wavelengths that anyone had ever taken of this object. The frosted-glass effect for radiation decreases quadratically as the wavelengths get shorter, and it looked like it would finally be possible to make out something more than just a boring splotch. Now the splotch was showing a small bulge in one direction. Were we seeing the blurry outlines of a small plasma jet, like we'd seen coming from the large quasars?

But the highlight of the miniconference were the spectacular results produced by the group led by Reinhard Genzel at the Max Planck Institute for Extraterrestrial Physics in Garching, just outside of Munich. Together with Andreas Eckart, Genzel had pointed a near-infrared camera at the Galactic Center. We use this kind of camera for night vision because it allows us to see the thermal radiation invisible to the human eye. This light has a much longer wavelength than the light that our eyes see, but as a result it also has an easier time piercing the dusty veil of our Galaxy. Suddenly there was something visible in the darkness of the center—a bright light. Was this the glow of the black hole?

This point of light was very blurry and out of focus because the Earth's atmosphere distorts starlight. When the light of the stars passes through the layers of air in the atmosphere after its long journey through space, it begins to flicker. We know this effect from hot summer days when the air rising from the hot pavement starts to shimmer and form schlieren that distort everything we see through them. The distortion in the atmosphere plays out in the same way at a larger scale. On Earth we have the impression that the stars are twinkling. Seen from space they don't twinkle at all, however. This is why space telescopes are so important for research. Still, the distortion effect on Earth isn't as bad for near-infrared light as it is for visual light.

Genzel and Eckart had come up with a trick for obtaining clear images from Earth. Instead of using a long exposure, they took a slow-motion film of the Galactic Center. With it they were able to capture the wild dance of this spot of light. In every individual frame the star looked frozen, and in the computer they could then correct for its leaping this way and that by cleverly layering all the images on top of each other. The near-infrared splotch became more and more clear and resolved into twenty-five different stars. So the light wasn't coming from a black hole. Still, one of these weak points of light was very close to radio source Sgr A*. Was this the long-sought-after counterpart to the radio source? We were all excited.

Astronomers have been hunting for the black hole at various wave bands for a long time, but again and again, the object they'd taken for Sagittarius A* would turn out to be just a star. Years later that turned out to be the case in this instance as well. If Sagittarius A* was a black hole, it was really very dark at almost every wavelength—except that of radio-frequency light.

Even if a lot of what was presented that day was just speculation and proved not to be quite accurate, I still had the profound feeling that something special was happening here, and that a new door to black holes was opening. Right now it was as if we were only seeing through a glass, darkly, but then face-to-face.

AN EARLY SUSPICION

After seeing the new VLBI images of the Galactic Center, my professor asked my fellow student Karl Mannheim, who later became a professor in Würzburg, and me if we couldn't explain the Galactic Center using a jet, like with a quasar. "It shouldn't take you but one or two weeks," he added with a playful grin. The subject seemed of burning interest to my well-traveled thesis advisor, and even Peter Strittmatter was suddenly listening attentively to what I had to say when he came back from Arizona.

And so I put the disk wind of quasars to the side and threw myself into work on Sagittarius A*. The two weeks turned into thirty years, and I'm still not done yet. I would never again work on the original topic of my dissertation.

What is Sagittarius A*? That was now the big question. What makes it shine? Is it really a black hole, a miniquasar? But Sagittarius A* was just a weak glimmer! If you placed quasar 3C 273 in our Milky Way, its center would be forty billion times brighter than what we were measuring there now. Could you even compare the two objects?

We used a simple model that Roger Blandford, one of the world's leading theoretical astrophysicists, had developed in 1979 with his doctoral student Arieh Königl to describe the radio emission of jets in quasars—though we did also add the possibility of regulating the power of these plasma jets. We fitted the quasar model with a gas pedal, so to speak.

You can picture cosmic jets as being like the jet engines on an airplane: hot gas is accelerated and shoots out of the engine at high speed. The more the pilot leans on the throttle, the more energy intensive, loud, and bright these engines get. In our quasar model, strong magnetic fields formed the engine; the energy was determined by how much matter fell into the black hole. If we took just about

10 percent of the energy caused by the matter falling in and stuck it into the magnetic fields and jets, we could explain the bright radio emission of quasars. Since black holes are relatively simple creatures, we didn't see why Sagittarius A* should be any different, fundamentally speaking, than its much brighter siblings.

Quasars gobble up approximately one sun per year. If our black hole sucked up ten million times less mass, the energy would still be completely sufficient to produce the radio emission of Sagittarius A*. Our Galactic Center would thus be a black hole on a starvation diet—though fasting would be a strange analogy, since ten million times less mass would still be three moons per year that it was devouring. A small stellar black hole, of which there are hundreds of millions in the Milky Way, would almost choke on that amount.[3]

We were also able to explain the size of the radio source, because thanks to its minimal power the radio plasma was no bigger than Krichbaum's VLBI measurements allowed. The radio jet would fit inside the Earth's orbit—compared to quasars, a real pip-squeak. No wonder you couldn't see it very well at a distance of 27,000 light-years.

Simultaneously with Krichbaum's VLBI observations we submitted our theoretical work to the academic journal *Astronomy & Astrophysics*. But something else occurred to me, something strange. In our model the radio emission shone like a rainbow—various colors along the radio light spectrum emerged from the center at various distances. The model predicted that as the wavelength got smaller the radio emission would get ever closer to the black hole. At a wavelength of seven millimeters, which Krichbaum had just used for his measurements, the plasma was still one astronomical unit, i. e. one Earth-Sun distance, away from the black hole, but in the one-millimeter wave band and shorter the plasma should have been coming directly from the vicinity of the event horizon. Put in terms of the rainbow, this would be the violet light of radio-frequency radiation from the innermost ring.

So did the millimeter-wave radiation that Mezger and Zylka had found come directly from the event horizon? The fact that the radiation seemed to cut off at even shorter wavelengths spoke in favor of this. Did the gas there no longer give off light because it had already vanished behind the event horizon?

I voiced my suspicion to Krichbaum and asked the question whether it might be possible to conduct a VLBI experiment at these frequencies in order to see the event horizon. Smiling, he said: "Yes, we're planning to of course, but unfortunately the Earth isn't big enough."

In 1979, the Max Planck Society, together with the Centre Nacional de la Recherche Scientifique in France and the Instituto Geográfico Nacional in Spain, had founded a new institute, the Institut de Radioastronomie Millimétrique (IRAM) in Grenoble. It operated two new millimeter-wave telescopes in Spain, and Bonn's Max Planck Institute built a third telescope in Arizona in collaboration with the university there. The radio antennas might conceivably be linked together for a VLBI experiment, but that was still too few telescopes to generate images. Plus, Krichbaum said, the black hole at the center of the Milky Way was much too small, as were all the others. Even a telescope as big as the Earth itself wouldn't be able to see the event horizon clearly enough at that wavelength. "Shame," I thought, but the idea was to keep nagging at me for a long time after that—and would never completely lose its hold over me.

THE SILENT MAJORITY

My doctoral thesis was comprised of five different articles for academic journals, and in the summer of 1994, after two frenzied years, it lay finished before me with the title "Starved Holes and Active Nuclei." Yes, "starved black holes," because contrary to popular belief, black holes generally aren't wildly voracious monsters: they're very well behaved and eat only what they're served. In our imagination, black holes might be giant, but compared to an entire galaxy they're just little chicks. And like chicks in the nest, black holes must wait for food, must wait for their mother galaxy to feed them with dust and stars. If this doesn't happen, they waste away, go dark and quiet, and stop growing—just like Sagittarius A*. But they don't die.

In my dissertation we had developed and supported the thesis that the compact radio emission of black holes everywhere follows the same principle: that it is the radiation of hot gas that gets spit out from the innermost edge of the accretion disk by magnetic fields in the form of jets. The jets flowing away from and the gas falling into the black hole from the accretion disk are closely linked, indeed almost symbiotic. There should be a universal connecting constant to account for the accretion disk and what's shot out in the jet. Put simply: the less that falls in, the less that comes out.[4]

In radio images, black holes look like fire-spitting dragons. Some are powerful and produce giant jets of far-reaching flame; others are weak and unenergetic, and from their throat escapes only a feeble wisp. But almost all of them produce jets: in this respect, the extroverted gluttons, the quasars, are no different than the fasting hermits of our Milky Way and its neighbors. Yes, even the radio emission of small stellar black holes can be explained using these jets. It's simply important that you concentrate on the radiation directly surrounding the gullet and not get distracted by the giant fiery plasma jets. You have to know exactly where to look.

Finally, my dissertation stated that the very same physics were at work in quasars, stellar black holes, and the Galactic Center. Or, to put it more in physics terms: black holes are scale invariant, and in principle, whether small or large, they always look the same in the vicinity of the event horizon. Black holes, it turns out, are incredibly boring. They have no hair, no neuroses, and no pimples. Why then shouldn't what goes on in their immediate vicinity look the same in every instance—at least when you're looking right down their throat?[5]

Most black holes don't particularly call attention to themselves. I once called them the "silent majority," because they behave like most people: only a select few really come out of their shells, become eccentric superstars, lead an exciting life, and have everybody looking at them. And so in the '90s, after all the hype around quasars, the focus around black holes turned toward the cosmic average population—even in the media. At the forefront of this shift was the Hubble Space Telescope.

The space telescope had cost many billions of dollars, was launched into space in 1990, and made only negative headlines at first because its mirror had been ground wrong. In a dramatic rescue mission, astronauts installed corrective optics in the space observatory. Now the telescope could look into the heart of our neighbor galaxies with a previously unknown level of clarity, and its measurements confirmed the tantalizing hints seen by more earthbound telescopes: stars in other galaxies also revolved at unusually high speeds around their Galactic Centers. Were there black holes at the centers of these galaxies, too?

Scientists cautiously called them "massive dark objects," or MDOs, but the well-oiled publicity machine at NASA regularly inundated us with press releases announcing that the Hubble Telescope had yet again—though always for the very first time—discovered a black hole in a galaxy. Later it would be water on Mars or an Earth-like planet that NASA had once more discovered—supposedly for the first time. Naturally, the Hubble Telescope hadn't located the black hole, but rather just the gas and the stars at a great distance surrounding the black hole.

In late May 1994 one of these first successful reports came in from NASA, and I was invited to the Westdeutscher Rundfunk radio studio to go on *Riff: Der Wellenbrecher*, a youth program, and talk about it. The live program was on the same day as my thesis defense, which meant that as a newly minted doctor of radio astronomy I had to hurry directly to the radio studio to arrive in time for the start of the show. The young moderator was a bit nervous, since she hadn't ever done an interview about physics before, and I had never given an interview live on the radio before. But the conversation went smoothly and was over before we knew it. We were both very relieved afterward.

The occasion for the interview was the Hubble Telescope's observations of M87. This galaxy was one of the "nebulae" that Charles Messier had spotted from the Hôtel de Cluny in Paris. Heber Curtis, the defender of the island universes theory, had seen a strange bright line that pointed outward from the center of the galaxy. Radio telescopes found out in the '70s and '80s that this line was a plasma jet traveling at almost the speed of light, exactly like the jets that had been found in quasars and radio galaxies—albeit significantly fainter.

During the radio interview I related how the Hubble Space Telescope had found out that at the center of M87 an unimaginable two billion solar masses had been compressed and balled together—probably a black hole. It was a thousand times heavier than the central black hole in our Milky Way. The moderator was a bit stunned, and even to me the number seemed unbelievably large. Well, sure, Americans like to exaggerate their findings a bit—that's probably the case here, too, I thought, but it's definitely something really big.

True, given its larger mass the black hole in M87 would be a thousand times bigger than Sagittarius A*, but since the galaxy was also two thousand times farther away, the event horizon would seem twice as small to us than the one in the center of the Milky Way—and even that was too small. "It's a shame, really," I thought, "but a miss is as good as a mile," because M87 also had a bright, compact radio core that you could have observed well even at short wavelengths.

If you want to see black holes, you have to shine a light on their surroundings. So it would surely be useful to understand where the light comes from and what kind of light is best suited for the purpose. Suddenly, however, there was intense argument about where the radio emission of starved black holes really came from. The American astrophysicist Ramesh Narayan at Harvard University was investigating what black holes that don't overeat look like. Unlike quasars, he claimed, a large part of the energy in them wouldn't even be radiated in the first place, but would disappear almost unnoticed into the black hole along with the intensely heated-up gas.

On this point I agreed eventually that Narayan was right. On another point, however, we had widely divergent views. In his model, the radio emission in the Galactic Center was supposed to come from gas in the accretion disk, shortly before it disappeared into the black hole. In our model the radio emission came from material that was just able to escape from the edge of the black hole by way of the jet. In M87 we could even see the jet directly in radio images; why should our Galactic Center work differently? Our model was meant to fit all black holes.

A dispute between unequal opponents: in one corner the renowned Harvard professor and in the other the young PhD student. Thankfully, conference organizers love nothing more than a nice academic feud, and I was invited to discuss the topic again and again. But which one of us was right? How to settle the argument? One thing was clear: we needed new radio data—above all from other starving giants!

Regrettably, only minimal or antiquated radio data were available. And so, little by little, I began carrying out observations myself in order to test my model. I submitted applications to conduct observations at the Very Large Array in New Mexico, the Very Long

Baseline Array in the US, and our own telescope in Effelsberg, and went hunting for black holes in other galaxies. This was a completely different sort of work than my theoretical calculations, but no less captivating.

The first time I was permitted to listen in on outer space with the 100-meter telescope in Effelsberg in the Eifel mountains, when I got to steer the giant white dish toward the preprogrammed celestial coordinates with the mere press of a button—it was a sublime feeling. Three thousand tons of steel were at my fingertips. I marveled, wide-eyed, at this spectacle of science and technology and felt like the little boy who finally got to ride on the great celestial garbage truck. In this instant it was immediately clear to me that I didn't want to just sit at my desk developing my theories; I wanted to experiment and test these theories and models myself.

I moved with my family to the US, spending two wonderful years in the peaceful little town of Laurel, Maryland. There, at the nearby University of Maryland and at the Space Telescope Science Institute in Baltimore, with the Hubble Space Telescope and other radio telescopes, I would sniff out and hunt down black holes.

THE DANCE OF THE STARS
AROUND THE BLACK HOLE

In Europe, the group led by Reinhard Genzel using the European Southern Observatory's (ESO's) telescopes in Chile sounded the call: the hunt for Sgr A* was on—first with a 3.6-meter telescope and later with the Very Large Telescope (VLT), an 8-meter telescope. But Genzel wasn't alone for long. A saga began to unfold, with two teams of researchers competing with each other, vying for supremacy in the center of the Milky Way.

The first showdown came in 1996, at a conference organized around the Galactic Center in La Serena, Chile.[6] I had presented a paper on Sagittarius A* and how much its radio emission resembled that of black holes in other galaxies. But the really exciting results came from Genzel's group. Their high-resolution images, taken over several years, now showed that the stars in the Galactic Center had shifted! If that was true, they had to be moving at breakneck speed.

We're used to the stars in the sky almost always looking the same, but that's not really the case, because in reality all the stars are zipping through the Milky Way at tens of thousands of kilometers per second relative to one another. Because they're so far away, though, it's not noticeable to almost anyone over the course of their own life.

The positions of the stars orbiting Sagittarius A* had changed within a few years—and this even though they're much farther away than the stars we see near us. Something or other had to be keeping these stars in motion and whizzing past each other. Only the gravity of a black hole weighing about 2.5 million times more than the sun could produce this effect, argued Genzel.[7]

The shift shown on the images was only minimal. A bit later, Andrea Ghez presented her group's results.[8] Ghez was a young professor at the University of California in Los Angeles and as of recently had one of the two 10-meter Keck Telescopes on Mauna Kea in Hawaii at her disposal. Her telescope was bigger and her observations promised to be even better, but because she had started later, she couldn't yet measure any shift. We'd have to wait a few years for that, but one thing was certain—a real competition was emerging here. In the following, the two groups eyed each other suspiciously and kept their data to themselves. At a later conference the two finally went up to the podium and reluctantly laid the transparencies containing their images one on top of the other. The measurements seemed to match up. For us this was very reassuring.

The La Serena conference was earth-shaking in another respect as well. One day there was a loud bang and the ceiling of the conference hall started quivering ominously. It felt like being punched in the stomach. Some of the attendees ran outside, afraid the building might collapse. It was the first earthquake I'd ever experienced. For the Chileans it brought back memories of earlier earthquakes that had caused the deaths of many people. Only the hardened Californians kept their seats. It's scarcely imaginable what would have happened if it hadn't been limited to this one tremor.

For me it was clear after the conference that something exciting was going to happen here. The race was on for new discoveries, and it was a race that was to last more than two decades. Science needs checks and balances. Competition is one way of making sure that these checks and balances are actually put into place. Like a pressure cooker, it accelerates development; it ensures that groups check each other's work, but it also creates enormous physical and psychological pressure. Competition functions when the competing groups are more or less on the same level. What's needed are good nerves, good health, sufficient financing, and functioning infrastructure over a period of many years. This was the case here, and it moved our understanding of the dark forces in the center of galaxies many huge steps forward. Where, if not here, were we going to be able to grasp whether black holes actually existed, and where, if not here, were we going to be able to sniff them out?

Three years later Andrea Ghez presented her next measurements of the stars' movements, and two years after that she was the first astronomer to discover that the stars moved in curved orbits.[9]

But where were they headed? All the stars seemed to be moving around a single point. But at the precise spot where this point was located there was—nothing. Sagittarius A* was still not visible on the images. Only a precise comparison of the near-infrared data with radio measurements taken by Karl Menten in Bonn and Marc Reid at the Smithsonian Astrophysical Observatory (SAO) in Boston showed that the point anchoring all movements was indeed the radio source Sagittarius A*.[10] Everything was spinning at a speed of several million kilometers per hour around the ominous radio source that sat motionless in the center.[11] Now it was clear: if there was a black hole there, it just had to be hidden somewhere in the radio light of Sagittarius A*!

Andrea Ghez also pointed out that one of the stars was traveling in a very tight orbit of just fifteen years around Sagittarius A*. Soon it would be time for it to make another pass, and it would be getting particularly close to the potential black hole, she informed us.

After that it was Reinhard Genzel's turn again, and he had installed a new infrared camera at the ESO's VLT in the Atacama Desert. This dusty, sparsely populated region in Chile is among the least hospitable you can imagine. The futuristic ESO hotel where most astronomers live looks like a supervillain's secret lair, and in fact the hotel appears in the James Bond movie *Quantum of Solace*. With their new array of instruments, the astronomers were able to take the clearest images to date of the center of the Milky Way. They used adaptive optics in which a deformable mirror corrects for the interference of the atmosphere within fractions of a second. The astronomers measured the star S2 and were rewarded for their efforts.[12] A comparison with old images showed that in only a few years it had moved to within just seventeen light-hours of Sagittarius A*. That's only about three times farther away than Pluto is from our sun.

This star was moving in an elliptical orbit around the powerful radio source, just like the planetary orbits around the sun that Kepler had described. And just as the sun and moon pull the Earth's oceans this way and that to create the tides, so does a black hole pull on the hot oceans of gas of a passing star. In this case, the tidal force of

Sagittarius A* wasn't strong enough to tear the star apart. This would only happen if it were to come within a bit less than 13 light-minutes of the black hole. But even so, the force of Sagittarius A*'s gravity on the little star is relentless. In its orbit it reaches the unbelievable speed of more than 7,500 kilometers per second—in a single hour it can cover a distance of 27 million kilometers. With the help of the old laws drawn up by Kepler and Newton, you can calculate the mass of Sagittarius A* based on the star's speed and distance and get a result of 3.7 million solar masses. This time the calculation came in higher than previous assessments. My heart leaped for joy, because it meant that the event horizon would also be larger and easier to see—but the margin of error for the measurements was still very large; the mass could vary by 1.5 million solar masses either way.

It had taken a good thirty years to go from Donald Lynden-Bell and Martin Rees's predictions in the 1970s to these measurements. Now that the scientific community was getting a glimpse of the dance of the stars around the suspected black hole, they slowly started to believe what was going on way out there in space. This black hole became the top celebrity of the Milky Way, and astronomers became the paparazzi, excitedly reporting on Sagittarius A*'s every movement.

Around this time Andrea Ghez's team discovered a star that orbited even a little bit closer to the Galactic Center. It took less than twelve years to circle the center of the Milky Way, and sped along on its orbit at one-hundredth the speed of light.[13] With their near-infrared telescope, Genzel's group finally managed to pick up a weak flickering in the precise spot where the radio source was located.[14] Now we could detect Sagittarius A* not just in radio-frequency light, but also in almost visible near infrared. X-ray telescopes in space also began to measure the flickering at the edge of the darkness.[15] Within minutes the emission would get brighter, then darker again. The emission could only be coming from a region with a width of just one light-minute—meaning it wasn't much bigger than the event horizon. Witnessing this cosmic spectacle, I thought that it was as if the black hole were covered in a rumbling storm cloud that kept sending out flashes of lightning. But the resolution of a single telescope was insufficient to pinpoint exactly what was going on there.

And so the Genzel group at the Max Planck Institute for Extraterrestrial Physics in Garching, along with colleagues in France and Germany under the leadership of the brilliant instrument builder Frank Eisenhauer, began to put together a technological monster program for optical telescopes, one of the most difficult and complicated of its kind. It was called GRAVITY and was meant to allow Genzel and company to link together all four of the giant 8-meter telescopes on the mountain in Chile, instead of using just one. In 2016, they finally succeeded.

During a visit to Munich in late 2017, I was able to see for the first time with my own eyes how the star S2 moves forward day by day. An unbelievably impressive spectacle for an astronomer! The data bore out the conclusion that the mass of Sagittarius A* was in fact 4 million solar masses. The margin of error was now less than 1 percent. Let's think about that for a moment: we now know the mass of the black hole in the middle of our Milky Way with greater accuracy than most of us know our own weight!

Since then, the GRAVITY team has regularly produced images that extend almost to the event horizon of the black hole and make visible the fascinating flares of radiation emitted by Sagittarius A*. The hot gas with its lightning flashes seems to reach almost the speed of light and to spin like a dizzying carousel around some object, consistent with what you would expect to happen in the vicinity of a black hole.[16]

Four hundred years ago we discovered that our planets orbit the sun. One hundred years ago we found out that the sun circles the center of the Milky Way. Ten years ago we saw stars that orbited Sagittarius A* like planets, and now we see gas 27,000 light-years away that rotates at almost the speed of light around a black hole. Again and again, gravity places celestial bodies and clouds of gas under its influence and forces them into their unchanging elliptical orbits. What a remarkable journey into the depths of the universe! No wonder that Andrea Ghez and Reinhard Genzel both received the Nobel prize for physics in 2020 for the discovery of a dark mass in the center of our Galaxy.

But the invisible object in the center: Was it really a black hole? We were so close to this mysterious object, and yet we were still denied the ultimate glimpse into this seemingly eternal abyss. We needed an even bigger telescope.

In the late '90s, during my stay in the US, I was offered the opportunity to get more deeply involved in the Hubble Space Telescope project and its possible successor. But I wanted to work in radio astronomy, and in 1997, I went back home with my family, hoping for the best. Anton Zensus had just started as the new director of the Max Planck Institute in Bonn, specifically to lead the VLBI group, and he spontaneously offered me a job. This was where the biggest global telescopes were made.

In Bonn in 1999, I also met my colleagues Geoff Bower, Sera Markoff, and Feng Yuan. Geoff had gotten his PhD at Berkeley and was a VLBI expert. We took a close look at the radio qualities of the Galactic Center and were later able to demonstrate, among other things, that in fact this black hole was hardly ever fed.[17] Sera was a theoretician and had gotten her doctorate in Arizona. Together we combined the radio emission of smaller and bigger black holes into a single model.[18] With my Chinese colleague Feng Yuan we took Ramesh Narayan's idea of a hot disk and tied it in with our jet model.[19] So began a productive collaboration that would last for many years, and I had the feeling that we were really starting to understand the fundamental principles of the astrophysics of starved black holes, both great and small.

THE IDEA BEHIND THE IMAGE

AMAZING GRACE

By the mid-1990s the net was slowly drawing tighter around our quarry, but there were still holes in it. To put it in legal terms, in our effort to prove that black holes were wreaking havoc in the centers of galaxies, we had so far relied on circumstantial evidence. As is so often the case in science, however, the evidence was by no means sufficient. You continue to gather facts in support of your hypothesis until at some point either no other conclusion is possible or the hypothesis has been disproved. Consequently, many astronomers remained skeptical—especially those of the old guard, who had seen their fair share of hype. "There's not enough evidence," they said. "We're still way too far away." Again and again, articles would appear that claimed that supermassive black holes couldn't possibly exist. Ideally, what we astronomers wanted was to catch the suspect in the act—and snap a photo that showed it still holding its prey in its hands.

What I wanted was certainty! I wanted to see black holes! More than anything!

The longing to see what is hidden must be an innate human need, something that is anchored deep within us. As a scientist I only believe what I can see, but first I have to believe that I'll eventually see it.

This longing to see takes hold of my soul again and again when I hear the old hymn "Amazing Grace." There's only a handful of songs that move me as deeply as this one, and there's one verse in particular that often brings tears to my eyes: "I once was lost but now am found, was blind but now I see."

This moment when our eyes open, when we suddenly grasp the truth, is of inestimable value—to emerge from darkness into light and be graced with the ability to recognize a new truth is one of the most precious experiences in our lives. Sometimes I think that this moment of revelation, when I think, "Finally, I can see!," is really the driving moment for me, the moment I live for. It's knowing that moment is somewhere in the future up ahead that gives me strength and spurs me on in the here and now.

That's probably what it all comes down to, in faith and in science: maintaining hope that you'll be permitted to discover something new. "Blessed are they that have not seen, and yet have believed."[1] This is how Jesus expressed this attitude toward faith, though I've always understood Jesus's words to mean something more akin to: "Blessed are they that have not *yet* seen . . ."

In everyday life one sometimes sees better with the heart; in science, however, we need instruments—big instruments. Today the highest-resolution images in astronomy are produced with Very Long Baseline Interferometry (VLBI), that is, the method that my colleague Thomas Krichbaum in Bonn, myself, and many other radio astronomers along with us have been using for decades now.

Starting in the 1960s, scientists began linking together individual radio telescopes to form an interferometer in order to increase the resolution of images. This method suddenly made details visible that no single telescope could ever capture. The result is a giant telescope with a virtual antenna as big as the Earth. Using this virtual antenna, the radio waves can be stored in the computer and then combined.

In order to combine the radio signals in such a way that the phases sync up perfectly, it's necessary to determine the position of the individual observatories to the nearest millimeter and to measure the arrival time of the signals with atomic clocks. These clocks function with picosecond precision; in 30,000 years they would only get off by a single second. The detected radio waves are converted into digital signals and transferred to a storage medium: in the past it was videotape; later, large reels of magnetic tape; and nowadays, hard drives by the crateful that store light saved as bits and bytes. The more data you can save, the more light you can capture simultaneously and the better the saved and secured data will be. The virtual telescope is assembled on the computer, and if there are enough data available, you can use algorithms to produce an image.

The measurements demand extreme precision, but also produce extremely sharp images. For that reason it isn't just astronomers who use continental interferometry to measure the heavens; geodesists also use VLBI telescopes to survey and measure the globe. We in turn also need the measurements from these land surveys, because the Earth isn't stable enough for our purposes. It deforms the virtual telescope, and geodesists measure this deformation.

Scientists at the Wettzell Observatory in Bavaria or the MIT Haystack Observatory near Boston, together with other stations around the world, regularly take a bearing on around three hundred quasars that are well suited to geodesic measurements. They are part of a worldwide geodesic network, and the data are correlated in Bonn or Haystack using the same methods astronomers use. Thus, both disciplines work closely together.

If you observe bright quasars like 3C 273 and 3C 279 and use them as reference sources, you can even use VLBI to correct atomic clocks and precisely measure the position of the telescopes you're using. In this manner, geodesists show how the surface of the Earth has changed. The distances between continental plates vary—for example, America and Europe drift a few centimeters apart each year. Hawaii, with all its telescopes on Mauna Kea, is the high-speed train among global observatories, careening toward Asia at a speed of almost 10 centimeters a year. Scandinavia has been rising since the end of the Ice Age thanks to the melting of ice sheets. Even the cathedral in Cologne wobbles up and down about 35 centimeters daily as a result of the tides. Thankfully this occurs evenly through-out the structure, otherwise the towers would have toppled on our heads long ago. Our global telescope wobbles!

The Earth's axis wobbles, too. The Earth is like a raw egg whose rotational axis is subject to tiny variations as a result of imbalances. The other planets tug at the Earth and cause the poles to wobble by hundreds of meters. The oceans do their part as well by flowing back and forth, as does the air mass that moves in the atmosphere around the Earth. As a result, the poles migrate unpredictably a few meters per year, and the precise deviation can't be predicted in advance. Many positions can be determined today by GPS, but the other planets rattle on the satellites as well. Our absolute position in space can only be measured by VLBI, and we need the exact po-sitions of the telescopes.

The image resolution[2] that can be reached with a VLBI network is calculated using this formula:

$$Image\ resolution = \lambda/D$$

The image resolution, that is, the pixel size of the image, expressed as a measure of angle, is the wavelength of the radio emission, λ (lambda), divided by D, the maximum distance between telescopes.

The smaller, i.e., better, the angular resolution, the smaller the objects you can distinguish. Observing at a wavelength of 1.3 millimeters and using the Earth's diameter of 12,700 kilometers as our baseline, the best resolution we can achieve is 20 micro arc seconds, which is about the size of half a mustard seed in New York that we try to observe from Cologne in Germany. If we calculate how big the event horizon of Sagittarius A* is given a mass of 2.5 million solar masses, as was assumed to be the mass at the time, then we find a diameter of 15 million kilometers. This is ten times larger than the sun. In the center of the Milky Way, however, this only appears to us to be about as big as a quarter of a mustard seed: 12 micro arc seconds—too small even for a world-spanning telescope.

And this estimation was still optimistic, I thought, because if a black hole is rotating at maximum speed—that is, at nearly the speed of light—then the event horizon will shrink by half. It was to be expected that every black hole rotates a little, just like every star and every planet. Would the visible part of the black hole get even smaller?

These were all the things I had to think about as I was sitting at the institute's library in Bonn one dreary afternoon in the mid-'90s. In my reading I suddenly stumbled on a short article by James Bardeen. The American astrophysicist had pondered back in 1973 what it would look like if a small black hole passed in front of a distant star. Back then this was a purely academic exercise, and actually still is today—because to see this cosmic constellation you would need an optical telescope at least 100 times the size of the Earth. Nevertheless, in my mind's eye I could already picture a black shadow passing over this distant sun, just like a Venus transit.

But something had me confused. The illustration at the end of the article showed a circle that was meant to indicate how big the dark spot would be that would result from light being swallowed up beyond the event horizon. This circle was far too big. Wasn't this a rotating black hole? Shouldn't it be much smaller? Five times smaller than the size shown?

The faster a black hole rotates, the closer light can get to it as it flies past. As if it were riding a carousel, it takes on momentum from the curvature of space-time and is just able to escape, while without the momentum it would be trapped from much farther out. For this very reason, I thought, rotating black holes should appear smaller. But this black hole appeared much larger to the observer—much larger than the event horizon.

And suddenly I got it: black holes enlarge themselves! They're gigantic gravitational lenses—because if there's one thing they can most definitely do, it's to deflect light. The black hole's rotation wasn't a problem either, since of course the light had to move past the black hole on both sides. True, on one side it flies past in the direction of the black hole's rotation and brushes right up against the event horizon, almost grazing it, but on the other side it has to fly against the current of space-time and is captured well outside of it. A black hole, then, casts a wide net to catch the light trying to fly past it.

It was as if the scales had fallen from my eyes. If what was drawn here was correct, and also held true for "my" black hole, then it would have to appear two and a half times larger than I had thought possible even in the best-case scenario. Whether it rotated or not made no difference as far as the observation was concerned—only the mass counted, and that we had a clear sense of.

In this case the Earth would be just big enough. Amazing grace! Maybe I could see "my" black hole after all. And not just me— everyone would be able to see it! The realization hit me like a flash of lightning. In my mind's eye a concrete image began to form. I now had a clear goal. I would stare down the throat of a black hole! I stood up, restless, and started pacing around.

THE BLACK HOLE CASTS ITS SHADOW

An idea that isn't shared is like a seed that's never planted. And so I went to conference after conference and started spreading the good news: "Yes, we can see the black hole." There would only be an image of a black hole if I managed to get colleagues from every corner of the world excited about such a project. It takes the will of many, all pursuing a common goal—but first they all have to be convinced.

Up to this point, though, it was all still theory. Theories are good. Theories that are supported by experiments are better. Experiments, however, only make sense when you can classify and explain their results with the help of a theory. Good experiments make theories better and stimulate new ideas, but they also cost a lot of money and effort. In order to raise the necessary budget, however, you need credible theories that predict what you're going to see: science is always a tango between theory and experiment, where first one leads, then the other.

So now we had to push forward with our telescopes to higher and higher frequencies, or shorter and shorter wavelengths. How close to the black hole could you get? In 1994, after the Bonn measurements at 7-millimeter wavelength, an American team at Haystack Observatory in Boston, including among others the young radio astronomer Shep Doeleman, conducted an initial VLBI experiment at 3-millimeter wavelength.[3] My colleague in Bonn Thomas Krichbaum even managed the first VLBI measurement at 230 GHz—1.3-millimeter wavelength—with the IRAM telescopes in Spain and France.[4] But what the object looked like we still couldn't say. Our Galaxy's frosted-glass effect still obscured its true structure, the quality of the data was bad, there were too few telescopes, and the sensitivity of the measurements was too low.

In 1996, I organized a coordinated observation campaign in order to observe the brightness of Sagittarius A* simultaneously with multiple telescopes at multiple wavelengths for the first time. Colleagues in Japan, Spain, and the United States joined in. We couldn't produce any images, but the interpretation of our data confirmed that the millimeter-wavelength emission really should be coming from the event horizon. In our article we made the explicit prediction that we should be able to see the event horizon against the background of this emission with a VLBI experiment,[5] but there was still great need for discussion among scientists worldwide.

The best place to hold discussions is at a conference, and so in 1998, my colleague Angela Cotera from Arizona and I organized a workshop on the Galactic Center.[6] Experts from all over the world came to Tucson. We deliberately chose a hotel in the middle of the desert so that nobody could run off at night and we would have lots of time to talk to each other.

At conferences the coffee breaks and group dinners are often more important than the presentations. "I'm not here for the presentations; I'm here for the drinks," an experienced colleague once told me, half joking. Humans are social creatures, and in eating and drinking together you learn a whole lot about one another and from one another that you won't find in any academic journal.

As planned, some heated debates sprang up. We didn't have lightsabers, but back then almost everyone had a laser pointer, which had just become affordable. At any given moment there might be three or four red dots dancing over the screen. This was all playing out in front of our guest of honor, Charlie Townes, who sat in the front row, the same Charlie Townes whose popular science articles on the black hole in the center of the Milky Way I had devoured as a student.

Did anyone notice the irony? Townes wasn't just anybody, after all. Here we were dueling with cheap laser pointers while there before us sat the man who in 1964, two years before I was born, had received the Nobel Prize for inventing the laser. But Charlie Townes, unlike us, still only used the traditional combination of finger and telescoping pointer! He seemed highly amused by the childish joy we took in his lasers. Any of us who took a moment to pause and think would have been struck by the amazing realization that, within a human lifetime, basic research had led to the creation of a common everyday object.

During the discussions, Krichbaum and I again emphasized that, using VLBI at high frequencies, we could reach the black hole and see its structure. My colleague Shep Doeleman, meanwhile, was still cautious and argued that the high frequencies could be coming from a dust cloud and not from the black hole. Suddenly Townes was wide awake. "Is there no hole in the middle of this thing?" he asked.[7] "That's right," I replied. "At higher resolution there literally will be a 'black hole' in the emission region that we could observe." Clearly we hadn't yet found the right term for this "thing."

Somehow my gospel of the possibility of seeing a black hole still lacked the power to get through to people. We had to do more. Humans like to have an image of things they're not able to clearly imagine so that they get a sense of what to expect. Up to this point I had only presented equations, graphics, and a schematic of a black hole. It was now time to show people exactly what we should be able to see—a simulated photo. To do so we would have to calculate the bending of light around a black hole and depict what it looked like when it was surrounded by a transparent, glowing fog, as would be the case in Narayan's accretion disk model or our jet model.

A few months later I received a stipend from the German Research Foundation, DFG, to go to Arizona as a guest professor for a few sabbatical months in 1999. Our youngest son had just been born and we made use of my wife's maternity leave. We arrived in Tucson with our three small children, Jana, Lukas, and Niklas, and just one of the eight suitcases we'd left with. When you spend a few days without much of anything, you take joy in the little things in life—especially your children.

My hosts put me in touch with Eric Agol, who at the time was on a postdoc fellowship at Johns Hopkins University in Baltimore. He had written a computer program that could be used to elegantly calculate the bending of light in accordance with the general theory of relativity—better than with the program I used for my master's thesis. Together we calculated what a black hole would look like under a wide variety of conditions, and whether you could see it with VLBI. Eagerly we waited for the results. And wouldn't you know it: in each of our models we could see a bright ring and a dark spot in the middle that was always the same size.

The conspicuous ring of light comes from all over. It's a result of the peculiar qualities of black holes: because of the warping of space, the light near the black hole flies in a nearly closed circle around it—so long as you aim past it at precisely the right distance. This closed orbit of light is called a photon orbit, because the light photons orbit like planets around the sun—but again, only at a precisely set distance. For a nonrotational black hole the photon orbit is one and a half times farther away from the mass center than the event horizon—but thanks to the gravitational lens effect it appears to us to be two and a half times bigger than the event horizon.

If a light bulb hangs above the black hole right along the photon orbit, then half its light falls into the black hole, the other half escapes, and a vanishingly small part of the light flies in a circle— namely the light that is emitted parallel to the event horizon. The closer the light bulb gets to the event horizon, the more of its light that gets swallowed and the less light that makes it out. What's more, the light gets stretched out and redshifted and loses energy. At the event horizon the light from the bulb then vanishes completely.

The space between the photon orbit and the event horizon is, in a manner of speaking, the twilight zone of a black hole; in this space, everything that falls in becomes rapidly darker.

In the vicinity of the photon orbit, the light can follow really crazy trajectories. Sometimes when I was a kid my friends and I would make supersecret spy telescopes with cardboard tubes and mirrors that you could use to look around a corner. A black hole is the ultimate supersecret spy telescope. It can look around several curves in every direction, all at the same time! With black holes you don't just have to be able to think laterally; you also have to be able to play all the angles!

If we had the ability to shoot lasers from our eyes like Superman, the trajectory of the laser beams would show us where we were looking. If, for example, we look to the left of the black hole, the sight beam will bend to the right and disappear around the corner. If we aim just a little bit farther to the right, the light will be bent a bit more drastically and come flying back around toward us—and we'll see what's in front of the black hole. Aim still farther to the right and first the light will be moving in a circle and then you'll be looking directly into the black hole. If we look to the right of the black hole, we see what's hanging out to the left, behind, or to the right of the black hole. If we look above the black hole, the light will be bent downward, and we'll see everything that's above, behind, and below the black hole. Indeed, the light in the vicinity of the photon orbit can fly in a quarter, half, or full circle around the black hole—sometimes even several tight, spiral-shape near circles.[8] Along the way more light is picked up.

If our gaze comes too close to the black hole, the sight beam ends at the event horizon and we look into darkness. We are literally in the dark as to the true nature of the black hole; bright light shines only in the area surrounding it.

If we fly around the black hole, we always see the same ring of light from all sides—in this sense the black hole is surrounded on all sides by a transparent light-emitting cloud. The light emission of this cloud is so bent and focused that it forms a thin, round, light-filled shell around the black hole. So from all directions we always notice a ring that contains a dark spot at its center. The darkness occurs because the sight beams end in the black hole. The spot isn't completely black, though, because the sight beams also have to travel through the luminous gas in the foreground.

The ring doesn't always have a uniform shape all the way around the spot, however. If we program a computer simulation to have the gas rotating at nearly the speed of light, as we would expect with a black hole, we only produce a half ring. On the side where the gas is moving toward us, the light is intensified, and on the other side it's weakened. What's more, if the black hole itself is also rotating, then the shadow and ring shrink by a few percentage points and even displays a small, scarcely noticeable flattening effect.

Twenty years later I would learn that the German mathematician David Hilbert had already worked out the mathematics of these light trajectories back in 1916[9]—only a few months after Einstein and Schwarzschild had laid down the foundations for black holes, without even knowing whether black holes existed and what they actually were. Hilbert's work was forgotten, probably because he was too far ahead of his time.

In the '70s and '90s, there were also a few efforts to calculate what black holes might look like,[10] but without a realistic chance of seeing them the work received scant attention. Only after our work was published did these papers slowly start to be rediscovered. With its release in 2014, the film *Interstellar* influenced how we picture them—though the model used for the film didn't really fit M87 or the Galactic Center. The black hole in the film isn't wrapped in a glowing hot gas cloud and has no jet; rather, it's surrounded by a thin opaque disk with a hole in the middle. It's not completely surprising that you see a hole in the disk when you've put one there beforehand. Even without a black hole a dark spot would be visible there. But it's only when it's actually supposed to be full of light that the darkness becomes meaningful.

When my two colleagues and I wrote our article forecasting the image of the black hole, we also talked about what we should call the "thing," the black spot in the middle. Evocative terms are important when you're talking about science. What would the Big Bang be without its bang? Everyone understands it, even though no one can really hear it. Dynamic terms are often able to convey abstract messages.

And so we scheduled a teleconference. We couldn't call it a black hole: that term included the mass in the center and the warping of space-time. *Hollow, spot, pimple*—somehow none of these words fit. Then all of a sudden the idea came to us to call it the *shadow* of the black hole.[11] We can't see any black hole directly, only its shadow, the missing light. The black hole hides behind its shadow and doesn't reveal all its secrets. A black hole is indeed just a shadow of its former self. This shadow isn't as crisp or as dark as a silhouette, either, because it's three-dimensional, and in this darkness you can always see a bit of light—from the gas in front of the black hole.

Naturally we wanted the simulated radio images we were presenting in our article to be impressive. How to illustrate something that we can't see with our eyes? It was clear of course that the image of the shadow of a black hole would only consist of data from a radio telescope. It wouldn't be a photograph in the classical sense, because our data didn't come from the wavelength range of light visible to the human eye. What color does such light have? We had calculated brightness levels, but not colors. In theory we could have used a contour or gray-scale image. This too would have meaningfully visualized the data—but it would have looked boring.

In the new millennium, the practice of using color images in astrophysical publications was becoming more and more accepted—though academic journals did require you to pay extra for color printing. To us it was worth it, though, because it was clear to me that the format of such an image would be crucial to its impact on readers. Back then radio astronomers liked to dip fairly deep into the virtual color palette and tended to choose rainbow colors for their graphic representations of celestial radio emitters—but a black hole isn't exactly a happy place.

A color scale called "Heat" struck me as far more appropriate. It represented molten iron. The shadow was now surrounded by a ring of fire, which was somehow reminiscent of the hot corona of a solar eclipse. I thought it was a very fitting color choice for the glowing monster that surrounds a black hole, but I was taking liberties for artistic effect.

In January 2000, we published the study in the *Astrophysical Journal* under the title "Viewing the Shadow of the Black Hole at the Galactic Center."[12] In it we described how it could be possible to see a black hole. It was a short "letter," which according to the journal's stipulations had to fit in just four pages—thus a few simulations appeared somewhat later in a conference volume.[13] Many of my colleagues still found the idea utopian, but nevertheless the short article became one of my most-cited papers. In a press release I proudly announced, "Soon we'll be able to see the black hole!"[14] In reality it would take another twenty years.

BUILDING A GLOBAL TELESCOPE

IN SEARCH OF TELESCOPES AND MONEY

Astronomy without telescopes is like a symphony orchestra without instruments. In order to take a simple picture with a global interferometer, you need at least five telescopes spread out across various locations; ten would be better. But short of stealing them, where could we get them? At the turn of the millennium there simply weren't enough telescopes of this kind, and what few did exist were threatened with closure for lack of funds. The long-planned construction of new telescopes kept encountering delays. For our ambitious project, things were more than a bit complicated.[1]

The 800-pound gorilla that was meant to dominate the field was the Atacama Large Millimeter Array (ALMA) in Chile, a global project that cost a billion euros and was being built by three regions working in collaboration: Europe, America, and Japan. This giant telescope was to consist of 66 individual antennas up to 12 meters in diameter—a network that would have the combined sensitivity of an 80-meter telescope and the image resolution of a 16-kilometer telescope. Already when we were writing our "shadow" article, it was clear that ALMA would be the key player in such a globe-spanning experiment. Doing VLBI with ALMA thus stood right at the top of our wish list,[2] and ALMA's own scientists were soon talking about the same thing.[3] But here, too, construction was delayed until 2011, and the VLBI capabilities were rationalized away: "We don't have the money to realize your project, but we'll make sure that it isn't made impossible," was the most hopeful response I received.

At my inaugural lecture at Radboud University Nijmegen in 2003, I talked about my dream of capturing an image of the black hole, and about how the more we learned about the universe the more we would come to recognize our own limitations. A Dutch newspaper ran a headline that claimed I was "rattling at the gates of hell."[4] I thought that had a nice ring to it.

In 2004 we got a small step closer to hell's gates. Geoff Bower, in collaboration with me and four other colleagues, managed to take the best VLBI measurements to date of the Galactic Center at long-millimeter wavelengths using the Very Long Baseline Array (VLBA).[5] The VLBA is a network of ten radio telescopes in the US—a continental telescope. The data were finally precise enough that we could calculate and compensate for the degree to which the hot gas in the Milky Way caused a loss of definition in the image. For the first time we saw the true size of the source as a function of wavelength, and just as our model predicted, it became smaller at shorter wavelengths, meaning that the shortest wavelengths should in fact reach the event horizon. Now it was finally clear that it really was the millimeter waves that were emitted in the immediate vicinity of the black hole. "After thirty years, thanks to radio telescopes, the fog has finally lifted," the Deutsche Presse-Agentur quoted me saying.

That same year the radio astronomers at Green Bank Observatory in West Virginia celebrated an anniversary.[6] The first sign of Sagittarius A* had been found here thirty years earlier, in 1974. In a solemn ceremony a plaque was revealed that commemorated the discovery. That evening I put together a special impromptu event where the scientists in attendance gathered to hear Shep Doeleman, Geoff Bower, and me discuss the shadow of Sagittarius A* and what technology we could use to measure it. At the end I asked for a show of hands: Was the time ripe for such an undertaking, or was the uncertainty still too great? The response from the audience was unequivocal: a clear majority of the gathered experts now believed in the image of the black hole—now we just had to find a way to take it.

After the workshop I invited Doeleman and Bower to a series of teleconferences[7] so that we could move forward with the experiment together. A global collaboration would be necessary, I thought, a collaboration like the ones the particle physicists liked to organize. There was no use in trying to be the Lone Ranger. The work would be planned, executed, and published in collaboration with many different researchers; experiment, data analysis, and modeling would be integrated into a single project.

We had clearly formulated our scientific goal. We intended to conduct a focused experiment that would either substantiate or disprove our hypothesis. Just as particle physicists were searching for the Higgs boson, we were searching for the shadow of the black hole. Either the shadow was there or it wasn't. We just wanted to investigate one celestial object, but in order to do so we needed the entire world. But bringing the world together would take a bit more time.

The Massachusetts Institute of Technology's Haystack Observatory, idyllically situated in the woods outside of Boston, was a leading center for VLBI and had now begun developing new hardware meant to enable the simultaneous storage of a substantially larger amount of data. Shep Doeleman was there to drive the program forward. He had gotten his PhD at MIT and came to Bonn for a short time as a postdoctoral fellow, where we met briefly. After his return to the US, he went back to work at Haystack Observatory. With four telescopes in Hawaii, Arizona, and California, Doeleman had at least a small network at his disposal. Just like me, he wanted to conduct the first test experiments.

In the meantime I was working at the LOFAR radio telescope, first as the project scientist and later as chair of the board. I was gaining firsthand experience in how large-scale physics experiments and international collaboration are done. In addition, I continued to work on the Galactic Center and on a few VLBI experiments, but in the Netherlands I lacked access to millimeter-wave telescopes. I had to wait for ALMA.

The Doeleman group first continued working with the four telescopes in the three locations available to them. In 2006, they pointed all the dishes simultaneously at the Galactic Center. At first they failed, but in 2007, they took successful measurements in the 1.3-millimeter wavelength range, and a year later the group proudly presented their results.[8] There wasn't an image yet, but the astronomers involved were able to determine the size of Sagittarius A* at the shortest wavelengths with a great deal more precision than the Krichbaum experiment ten years earlier. Sagittarius A* was in fact exactly as large as one would expect given the shadow and its ring of light! Now the excitement was really mounting, and I was awfully pleased—again the theory had been confirmed. You just couldn't see the shadow yet!

Doeleman worked hard to drum up support in the US, and I tried to do the same on the other side of the Atlantic. In order to raise a large amount of money, you need a large amount of support from a wide range of sources. In 2007, for the first time, European astronomers came up with a joint strategy paper for the future of astronomy,[9] and our shadow experiment was included. Now our idea was officially recognized as one of the most important European scientific goals of the coming decade, and the same thing was to happen in the US. "Astro2010: The Astronomy and Astrophysics Decadal Survey," a ten-year program for the US, was put out under the evocative title "New Worlds, New Horizons in Astronomy and Astrophysics."

Shortly before the decadal survey was released, Doeleman organized a workshop at the annual meeting of the American Astronomical Society (AAS) in Long Beach, California, to which I was invited. The goal was to emphasize the far-reaching international support for the decadal survey.

During a coffee break I sat down with Doeleman and Dan Marrone, who at the time was in Chicago and who would later go to Arizona. In the past few years it had become ever clearer to me how indispensable good marketing is for an undertaking such as ours—even in science, this is the case. But here we didn't even have a memorable name for the project. Nobody aside from a few nerds would really know what to do with "Submillimeter VLBI Array." "This has to change, and quick! We need an attractive name," I told the group, and suggested the name "Event Horizon Array." After a lively discussion we agreed on "Event Horizon Telescope," EHT for short. A name, a symbol, a brand was born—in one of those famous coffee breaks in which you sometimes get farther and make more progress than in whole days' worth of lectures.

Later, some of the meeting's participants published their strategy paper for the decadal survey.[10] In it the project officially traded under its new name for the first time.

In America the money was now flowing a bit more liberally, and radio astronomers in Bonn were also continuing to get more involved in the new VLBI experiments with the IRAM telescopes in Spain and France, as well as the new Atacama Pathfinder Experiment (APEX) telescope in Chile. Then, in 2011, it was the Netherlands' turn. On a lovely early summer day I got a surprising phone call from Jos Engelen, former chief scientific officer at the European Council for Nuclear Research (CERN) and now head of the NWO, the Dutch Research Council. We knew each other through my work on astroparticle physics. "I hope you're sitting down," he began. Surprised, I stood up. "Dear Heino, I'm calling because I wanted to personally inform you that you have won this year's Spinoza Prize for your work with LOFAR and on the visualization of black holes," he said weightily. Granted, this sounded totally great, but what was the Spinoza Prize? As a foreigner, there were embarrassing gaps in my knowledge. Luckily, before I could ask, he explained, "This is basically the Dutch Nobel Prize!" For a second I wanted to ask him if that made sense, like saying the national Dutch world championship, but I kept my remark to myself. "It's significantly better endowed than the Nobel Prize," he continued. "You'll receive 2.5 million euros." Now I sat down. "You can use this prize money however you want—for research purposes, of course, not for personal use," he added. I knew immediately what I would use the money for.

BUILDING THE EVENT HORIZON TELESCOPE

A few months later, carrying the proverbial suitcase full of money, I traveled to the Event Horizon Telescope's first international strategy meeting in Tucson, Arizona. The large ALMA Telescope in Chile had finally been finished, and now the key representatives of the crucial research institutions and observatories would all be coming together. I saw many close colleagues.

At length we discussed the latest scientific findings, such as those in the realm of theory. The calculating capabilities of so-called supercomputers had made great advances in recent years. In the same way that they predicted the movements of air masses around the Earth for weather forecasts, these mammoth calculators could simulate how gas moved around a black hole. The method behind these simulations is known by the acronym GRMHD, for "general relativistic magnetohydrodynamics." This sounds complicated, and indeed it is. GRMHD simulations involve highly complex models that simulate magnetized plasma flows in warped, rotating space-time systems. Still more programs calculate how light and radio emissions are produced, bent, and absorbed by the hot gases surrounding a black hole. These computer calculations are substantially more extensive than anything we'd calculated in 2000. The megacomputers were now producing wonderful, captivating images, and with their computing power astronomers everywhere were finding shadows of black holes and in so doing confirming our fundamental hypotheses. A real "shadow industry" was emerging, and in almost all of the models, shadows and rings of light were visible—thus, at the level of theory, there was broad agreement.

A young scientist, Monika Mościbrodzka, impressed me with her competence and her attitude. She had gotten her PhD at the Nicolaus Copernicus Astronomical Center in Warsaw under Bożena Czerny, a well-known accretion disk theoretician, and had learned the tools of the trade under Charles Gammie, one of the leading experts in numerical simulations in the US. Now she had managed to produce one of the best "weather forecasts" for Sagittarius A*.[11] Up to this point, this area of research had been dominated exclusively by men, but Monika wanted to make her mark. I offered her a position in Nijmegen and asked her to put together a numerical simulations team. This is a tough row to hoe. Programming, carrying out, and analyzing the simulations cost extraordinary amounts of time and energy and demand the tenacity to be able to spend many lonely hours at the computer. Every publication is hard fought. It's as if you were to take the telescope with which we gather the data and reconstruct it, detail by detail, function by function, on the computer. Later Monika would manage to update our old jet models from the '90s[12] and generate an astoundingly accurate prediction of the image the EHT would eventually produce.[13]

Another much-discussed advance at the conference was the fact that the mass of Sagittarius A* was greater than previously thought, and that the black hole in M87 had also grown in the past few years. M87 was now thought to weigh not two billion, but rather three billion solar masses. One team of researchers even claimed that this black hole was six billion times heavier than the sun. If that was true, the shadow would have to be big enough that we could see it! Did we now have two candidates to work with? The black hole in M87 would still be a bit smaller than Sagittarius A*, but it was in the northern part of the sky and was easier to see from the northern hemisphere, where most of the telescopes were located. Plus in this case the Milky Way wasn't in the way. It wouldn't blur an image of the black hole in M87—we would have one problem fewer. Oh, this was probably too good to be true! I was cautious. Was this not a case of the wish being father to the thought? Systematic failures had too often been evident in attempts to determine the mass of black holes in other galaxies, but still, it would definitely be worth a try.

At the strategy meeting in Tucson, the scientists held their discussion in the conference room. Meanwhile, in the back room, the directors of observatories and of important institutions hashed out the scientific politics. I was right in the middle of it, and in the end we agreed on a common plan for moving forward. The foundation for a global course of action was laid.

Now it was serious, and more funds had to be raised. Acting on their own, neither the individual telescopes nor the large observatories, ESO and NRAO (National Radio Astronomy Observatory),

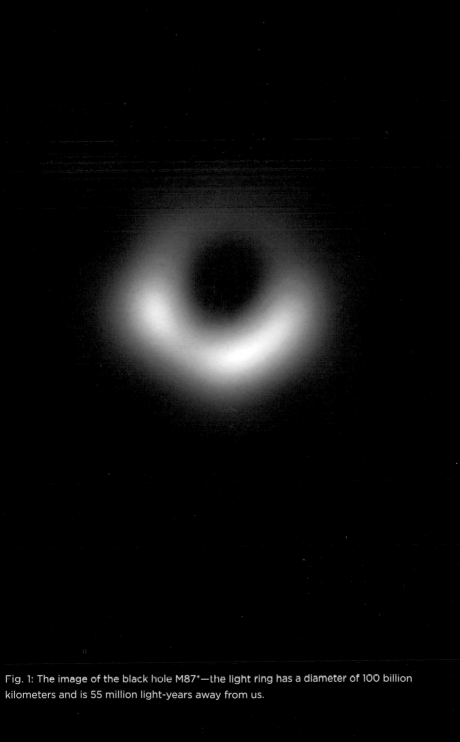

Fig. 1: The image of the black hole M87*—the light ring has a diameter of 100 billion kilometers and is 55 million light-years away from us.

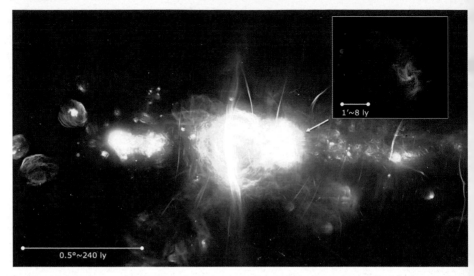

0.5°~240 ly

1'~8 ly

Fig. 2: The center of the Milky Way in radio-frequency light (MeerKAT, South Africa, and VLA, US). We are seeing the glow of hot gas and magnetic fields in the disk of our galaxy at a distance of 27,000 light-years. The bright spot to the right of center contains Sagittarius A*—our central black hole.

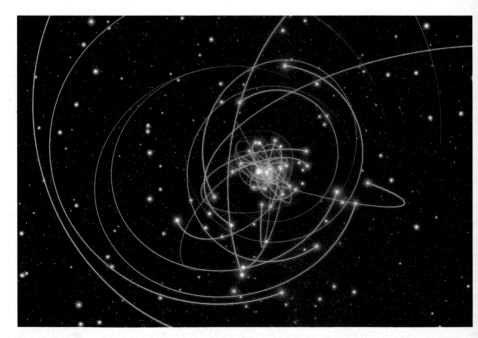

Fig. 3: A dance of the stars around the black hole in the Galactic Center—simulation based on actual measurements of the stars' movements. The stars speed along at a few thousand kilometers per second around a single point: the radio source Sagittarius A*.

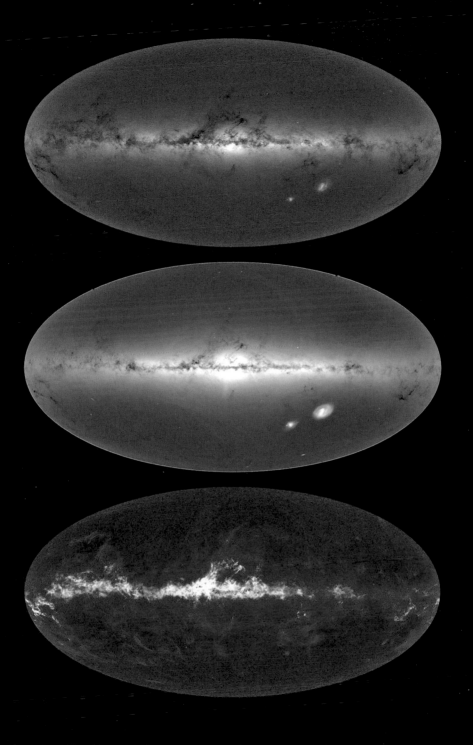

Fig. 4: Our Milky Way—1.7 billion stars, measured by the ESA's *Gaia* satellite: luminosity *(top)*, number *(middle)*, interstellar dust *(bottom)*. The entire sky is depicted.

Fig. 5: Dust ring around the protostar HL Tauri, 450 light-years away—a new solar system forms. The disk is about three times as large as Neptune's orbit around the sun. We are looking at millimeter-wave radiation that was captured by the ALMA Telescope.

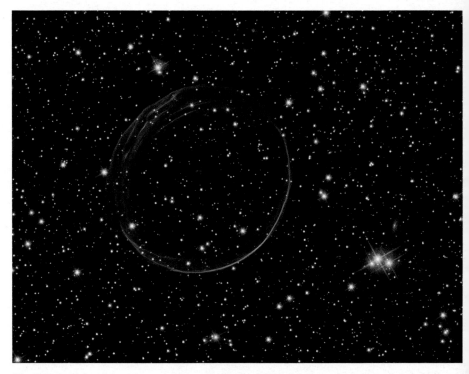

Fig. 6: Remnants of a supernova with a diameter of 23 light-years: the result of a stellar explosion, which creates at its center a compact neutron star or even a black hole.

Fig. 7: The Andromeda Galaxy—sister of the Milky Way. It is one hundred thousand light-years in diameter, consists of hundreds of billions of stars, and is about 2.5 million light-years away from us. In the brown dust clouds seen in the disk, new stars are formed.

Fig. 8: The large elliptical galaxy Hercules A is located in a galaxy cluster 2.1 billion light-years away. A black hole shoots a plasma jet out into space that is 1.6 million light-years in length. The red is from radio image captured by the VLA. The black-and-white and other color is from the Hubble Space Telescope.

Fig. 9: A detailed GRMHD computer simulation of a black hole: accretion disk in red, plasma jet in gray. The shadow of the black hole is visible in the center, where light disappears beyond the event horizon.

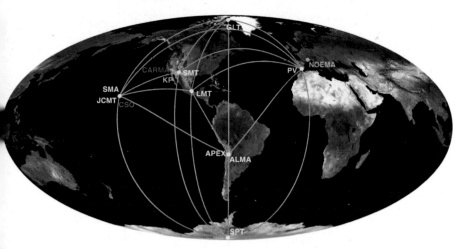

Fig. 10: The Event Horizon Telescope (EHT).

Fig. 11: One of the telescopes in the EHT network: the IRAM 30-meter telescope, Pico del Veleta, with the crew after the observation campaign of 2017 (*from left to right*: S. Sanchez, R. Azulay, I. Ruiz, H. Falcke, and T. Krichbaum).

Fig. 12: Another telescope in the EHT network: JCMT, Mauna Kea, 13,425 feet (4,092 meters) above sea level, Hawaii.

Fig. 13: Yet another telescope in the EHT network: ALMA, Chajnantor Plateau, 15,700 feet (4,800 meters) above sea level, Chile.

could finance the EHT project, carry out the science, or conduct the appropriate analysis. The observatories needed their resources and staff just to keep their telescopes in operation. It was up to us now; we had to spring into action. But what to do?

Sometimes chance helps. On the way back from a LOFAR meeting in Dwingeloo in 2012, I ran into my colleague Michael Kramer on the train. We had both gotten our PhDs at the same time, but our paths hadn't crossed since. By then Kramer had become the third director of the Max Planck Institute for Radio Astronomy in Bonn and had conducted fundamental and successful tests of Einstein's theory of relativity using pulsars. We quickly found a common wavelength. Five years earlier the two of us had gotten a lot of money from the European Research Council (ERC), becoming the first astronomers to do so. I had used it to finance the pioneering measurement of cosmic particles with LOFAR; he had built a VLBI-like network for the measurement of gravitational waves using pulsars. We were both fascinated by gravity. Our projects were winding down and our financing was running out, and we were both itching to start something new.

I talked about the EHT; he described how a pulsar can be used to measure the space-time around a black hole with unbelievable precision. We decided to submit an application to the ERC together and go up against the best research teams in Europe in every field—even though the chances of successfully obtaining the hefty sum of 15 million euros were only 1.5 percent.[14] To make it a three-member group we managed to win over Luciano Rezzolla, an Italian astronomer who had first worked on gravitational waves and melting black holes at the Albert Einstein Institute in Potsdam and now taught at Goethe University in Frankfurt.

We took some time to get to know each other and soon were moving full steam ahead. The three of us worked for half a year on our joint application and called the project BlackHoleCam.[15] Every telescope, including the EHT, needs a camera—and we wanted to provide this camera. For the Event Horizon Telescope the camera was the combination of the data recorders and the analysis software.

Then in the first months of waiting a small miracle took place. One element of our application had to do with using ALMA to search for pulsars in the Galactic Center, an extraordinarily risky and speculative undertaking. For decades astronomers had been on the lookout for pulsars in the center of the Milky Way. There should be thousands of them there—but not a single pulsar had been detected to date. As luck would have it, within the few months that our application lay before the evaluators, a completely new pulsar flared up in the Galactic Center for the first time. We were the first to discover and measure it with the 100-meter telescope in Effelsberg. In September 2013 the magazine *Nature* published our work,[16] and we received a great deal of attention. The result demonstrated that it was possible after all to find pulsars in the direct vicinity of the large black hole in our home Galaxy. Nature had done us a huge favor, because this discovery certainly didn't hurt our application. Just how many pulsars were still hidden out there?

To our surprise, to this day no second pulsar has been found in the Galactic Center, despite intensive searching. Why that is remains one of the biggest puzzles of the Milky Way. Just as puzzling was why this pulsar decided to call attention to itself in the very months when we needed it—but we weren't making it up. Other astronomers confirmed our findings. Didn't I say once before that sometimes in science you just have to get lucky?

The selection process was like a casting show. Our application had to make its way through round after round, and at the end of each round a ruthless jury would give it a thumbs-up or thumbs-down. We actually made it to the final round and were invited to Brussels to appear before the jury in person. Now we really didn't want to fail.

We spent days rehearsing for our appearance, prepared ourselves for every possible question, and when the time came, traveled to ERC headquarters in the European capital.

The three of us entered the anteroom in the best of moods. The team that was to present before us was already there waiting. Here were highly respected professors from world-renowned Oxford University sitting hunched over or pacing nervously back and forth.

After twenty minutes another group came back from their interview—all the participants looked crushed. "They ask very specific questions about the financing plan!" one of them groaned. Our hearts started to sink. Some of the best and most experienced scientists in all of Europe were gathered here, and they all felt like schoolkids waiting to take an oral exam. When we walked into the room for our presentation we saw the twenty-person commission arrayed in a U shape before us. Like Roman gladiators we stepped into the arena, staring scientific death in the face. But where were the trumpets?

Our presentation couldn't have gone better. Michael, Luciano, and I passed the ball back and forth in perfect sync, and then we stuck the landing, ending within a second of the allotted time. Now the commissioners asked questions that we parried as a team, well rehearsed and in perfect harmony. The only astronomer on the commission, Catherine Cesarsky, was a former director of the ESO and knew exactly what she was talking about. "What is your relationship to the Event Horizon Telescope?" she asked, zeroing in on the weak point in our application, the still-vague organizational structure of the EHT. What if there was conflict and it all fell apart? "We intend to be a part of it and to play a role in putting it together as a way of combining resources," we said, "but we need money to be in a good position to negotiate. If necessary, however, we are also prepared to conduct the experiment on our own." Catherine Cesarsky smiled;

apparently we had given the right answer to one of the most important questions. We had the commission on our side.

Our time was almost up. "I have one more question." Now another member of the commission spoke. "I don't understand these two items in your budget concerning public outreach. Could you explain?" My heart pounded. He's asking about figures! My head was suddenly one big black hole. I stammered out some vague generalities. The interview ended; filled with uncertainty we went back home. Had the presentation gone well? Had it worked? Or had we failed in the last five minutes?

Two weeks later we got the letter from the president of the ERC. I've received lots of these letters in my life. You just have to read the first four words, then you know everything you need to know. Here I read: "I am very pleased . . ." The application was accepted! I stood up and walked around my study. I felt happy and calm. The commission had docked us a million euros because I'd floundered in the last five minutes—I'd never lost so much money in so little time before. Still—we'd done it! We were the first to bring real money to the table for the EHT—14 million euros. Would a successful cooperation with the Americans be possible now?

That same day I wrote an email to Doeleman and told him I wanted to meet him in Boston. I booked a flight and three days later was sitting in a room with him and the director of Haystack Observatory, Colin Lonsdale, who mediated in his calm and even-keeled way. For two days we discussed the next steps and agreed to sign an interim letter of intent declaring that we would work together on the Event Horizon Telescope.

Together with Dimitrios Psaltis, a gravitational theorist from Arizona, and other colleagues in the US, Doeleman was working on his own big application for the National Science Foundation (NSF), America's biggest funder of scientific projects. We wrote a letter of support, stating that we agreed to work with Doeleman's team. This application was also accepted, and the EHT was allotted 8 million dollars from the NSF headquarters in Virginia. Now we had enough money together and could get started on concrete plans for the new experiment.

The meeting in Boston wasn't the end of it. Back in Europe we had to put our team together. I managed to bag Remo Tilanus, an experienced astronomer, as project manager—he had headed up the James Clerk Maxwell Telescope in Hawaii on behalf of the Netherlands for years, until the Dutch bowed out in 2015, and had made significant contributions to the VLBI experiments there.

At the same time, in Nijmegen, five young grad students appeared at my doorstep unexpectedly who were interested in the EHT and whom I could offer spots in our PhD program—a fantastic troupe that was just beginning to come together. To me they seemed heaven-sent, though many of them were from the area around Nijmegen. Ultimately we ended up having seven nations represented on our team.[17]

"We're going to conquer the world, but in a nice way"—this was our motto, and I drummed it into everyone. My students and the people I work with enjoy a lot of freedom. My goal is for them to find out for themselves what it is that drives them. Ultimately each of us has to find his or her own place; only then can a person work on something with heart and soul. What's important is that everyone have their own goal to follow that suits their actual talents, and that these talents supplement those of the others in the group, rather than have everyone competing against one another.

At a workshop in November 2014 at the Perimeter Institute, a research institution for theoretical physics in Waterloo, Canada, the backroom diplomacy came to a head and the presentations were almost beside the point. Here we had a few dozen astronomers fighting for their roles in the EHT. Who should be a part of the leadership circle? What was the organization's structure? On the last night the knot had to be either untangled or cleaved in two. The negotiations lasted well into the night; a fight without fists, though at least one of us might have pounded their fist on the table and then, toward midnight, we had finally reached an agreement on what the EHT was to look like in the future. We sealed it with handshakes and agreed to leave it at that, but by the next morning a few of us were still trying to renegotiate things.

Only after another fifty teleconferences did the EHT become a provisional collaboration. This was in the summer of 2016. A year later all the paperwork was officially signed—and by then we had already conducted our experiment. The collaboration brought together thirteen institutional partners: four in Europe, four in the US, three in Asia, and one each in Mexico and Canada. Each was equally represented on a board that formed the top leadership body.

A three-person team consisting of a director, project manager, and project scientist was to assume responsibility over day-to-day management; an elected eleven-member science council decided on and supervised the scientific program.

Shep Doeleman became director, Dimitrios Psaltis became project scientist, and Remo Tilanus became operations manager. I was selected to be head of the science council and my colleague of many years Geoff Bower was selected vice-chairperson—he had moved to Hawaii by then and begun working for Taiwan's Academia Sinica Institute of Astronomy and Astrophysics. Anton Zensus, director of the VLBI team in Bonn, and Colin Lonsdale from Haystack Observatory assumed leadership of the board.

This was how the power was divided, but there wasn't a single woman in a leadership position or on the board. This is a problem that has plagued the EHT from its inception, and not something we can be proud of. Only on the science council were two women represented, one of them being Sera Markoff, who was now teaching in Amsterdam.

ON EXPEDITION TO ARIZONA

Alongside the intense and hard-nosed negotiations, we had long been busy preparing for our first expeditions. The ALMA Telescope was finally ready for initial VLBI measurements,[18] which were to take place in January 2015. Now we had to show that we could handle a major experiment, both technologically and organizationally. All the telescopes were to be fitted out with the same VLBI equipment, all of it latest generation.

On September 1, 2014, the first wave of funds comes in from the European Research Council. That very same day Remo Tilanus sends out the necessary orders for VLBI equipment so that the particularly important long lead items make it to the telescopes on time. So-called Mark 6 data recorders are ordered from a company in Boston. Technicians in Groningen build electronic filters at short notice with help from Bonn and blueprints from Haystack.

Hundreds of the latest hard drives are to be sent via Haystack Observatory to the different telescope stations. The order is delayed, and after a blizzard in winter 2015 all of New England is covered in a thick blanket of snow and ice—everything comes to a standstill. A colleague slips on the ice and suffers a compound fracture. We can't purchase hard drives in the large quantities we need in the US, and the dollars aren't flowing yet. Remo Tilanus has to improvise. In just five days he manages to place a large order for drives through the ordering system at Radboud University in Nijmegen, and then have them flown from the Netherlands to Boston. From there they can be distributed all over the world.

How he pulled this off, nobody knows to this day. These aren't just the wonders of globalization; they're also the uniquely heroic deeds of a project manager, deeds that hardly anyone notices. In the end all the necessary components make it to the telescope stations just in time and are installed and tested by the local technicians on site.

Everything is ready, and in late March 2015 we all fan out to our various destinations around the world for our first large joint expedition. We intend to link up as many telescopes around the globe as possible. I travel to the US, to the Submillimeter Telescope (SMT) atop Mount Graham in Arizona. I drive from Tucson through the adventurous landscape of the American Southwest, past bare rocks, cacti, small towns built of transportable wooden huts, the impossible-to-miss "The Thing" souvenir shop, and a desert prison whose prominent billboards warn of escaping prisoners. Taking a detour to stop in the small town of Safford, not far from the road up the mountain, I stock up on supplies for the coming week. If you're starting your journey through the Galaxy in Arizona, you have to bring everything yourself—aside from a towel.

At base camp at the foot of the mountain I get a security pass and a walkie-talkie. Then the adventure really begins with the drive up the mountain on Arizona State Route 366. I had to rent a car with four-wheel drive at the airport and was given a giant red Dodge Ram pickup truck to have all to myself—this feels very American, low gas mileage included. Up I go, from 1,000-meter elevation in the high plains to the 3,200-meter peak of Mount Graham. There the telescope sits on a small plateau. On the way I see signs for the Arizona Church of Christ Bible Camp and the Shannon Campground. The landscape changes. At first it looked like the backdrop to a Western; now snow-covered peaks and forests of fir trees that wouldn't be out of place in the Pyrenees line the remainder of my drive.

The paved road ends. Past a barrier it continues, bumpily, until a fallen tree blocks my path. I've missed my turn and have to turn around. I'm already tired and this last stretch is especially steep and narrow, so narrow that only one car can drive up or down it at a time. I radio up to ask if the way is clear: "Is there any downhill traffic on the access road?" I ask, get no answer, announce myself— "One vehicle proceeding up the access road"—and step on the gas. The truck jounces up and down as it struggles up the switchbacks. The rough gravel road leads farther and farther up the mountain—I don't even want to think about what the road is like when it's raining. Suddenly I reach a wide parking lot—that indispensable American architectural feature—stretching out in front of the telescope a short ways below the summit.

The 135-ton structure is an impressive edifice. Sleeping quarters and a kitchen are located at the base. The upper, movable part of the building houses the telescope and the instruments. The front wall and the roof can be retracted, giving the 10-meter dish a clear view of the sky—if none of the trees, which are protected here, are in the way. Otherwise the telescope is snugly housed inside the building complex. A flight of stairs leads up to a small landing and the control room, located right below the dish. When the telescope changes position, both the control room and the stairs rotate along with the outer part of the building. This leads to constant confusion over the course of our observations as the stairs keep moving along with the telescope whenever we take a bearing on a new set of coordinates in the sky. Every time I leave the kitchen or my bedroom, I have to go looking for the stairs again to get back up—and they're always in a different spot. It's enough to drive you crazy!

The dish itself consists for the most part of a carbon fiber–reinforced plastic structure coated in a thin layer of aluminum. The antenna shines like a giant mirror and must never be pointed in the direction of the sun; otherwise it would turn into a giant magnifying glass and melt. One of the first submillimeter telescopes lost its antenna in this fashion.

The mountain is an ideal location for observing the sky because the concentration of water vapor in the atmosphere, which causes attenuation and corruption of the radio signal, is much weaker up here. For some people the thin air takes some getting used to; they get rather short of breath. I have a mild headache myself, but thankfully I'm able to get up the stairs to the control room well enough; all that soccer and volleyball playing is paying off after all. My throat is dry from the lack of moisture in the air and my skin starts peeling. As a result I wake up a lot during the night—such is the astronomer's lot. Even my supplies are affected by the low pressure: the bag of chips I bought is inflated, and the ball in my deodorant stick shoots out at me with a loud pop when I open it. It barely misses me, and the deodorant sprays all over the bathroom. Hopefully it doesn't get too hot in the coming days—I guess I'd better not sweat.

Outside, on the peak of Mount Graham, I'm surrounded by a fragrant fir forest. I reach a clearing and a sublime panorama opens up, the sparsely populated expanse below me and the sky up above. As a radio astronomer working in the submillimeter wave range, you hope for a cloudless sky so that the waves can reach the dishes through the atmosphere with as little impediment as possible. Normal radio waves penetrate the clouds easily, but the short waves that we want to observe get absorbed by the water vapor in the air and in the clouds.

Mount Graham is astronomer territory. Two hundred meters to the east of the SMT, a large gray hulk soars above the treetops. This is the Large Binocular Telescope—a giant optical telescope with two 8.4-meter mirrors in which German research institutes have a one-fourth stake. The mirrors, with their honeycomb technology, are a specialty of the University of Arizona's mirror lab. "You can have any mirror you want," former director of the Steward Observatory Peter Strittmatter told me before I drove up the mountain, "so long as it's exactly 8.4 meters wide." Strittmatter was very good at selling telescopes.

West of the SMT lies a small, nondescript but nevertheless just as special observatory. It houses the Vatican Advanced Technology Telescope (VATT). The long building is almost a bit reminiscent of a church. A long nave leads to a silver dome. What's below it isn't an altar, but rather an optical telescope 1.8 meters in diameter.

We can still feel the influence of the Vatican's astronomers today. It was they who developed our present-day calendar in the sixteenth century. The modern observatory was founded toward the end of the nineteenth century in Rome, but then moved to nearby Castel Gandolfo when nocturnal street lighting was introduced. In the twentieth century an affiliated research center was established in Arizona.

One night when we have no observations scheduled I pay the neighbors a visit. There are three Jesuits currently working at the Catholic observatory. They're searching for asteroids that could potentially pose a danger to Earth. I enjoy the calm, friendly atmosphere. Father Richard Boyle is here; he used to run a summer school at the Vatican for astronomy graduate students from all over the world in which I took part. These days he seems to spend nearly all his time behind the telescope and lives on the mountain almost like a hermit. Life in an observatory does in fact take on a somewhat monastic and meditative aspect. An astronomer conducting observations orders his life around the sky. Stars and galaxies set the rhythm. Nothing is there to distract you. I enjoy this time on the mountain, where life is simple and one is closer to the heavens.

Our team in Arizona consists of a handful of EHT colleagues. Among them are Vincent Fish from Haystack Observatory and Dan Marrone from Arizona. I take over for Dan up here and he continues to offer supervision from Tucson. There aren't enough beds in the observatory for such a big team, so not everyone can be here at the same time. Right from the outset, I feel at home at the SMT. Of course I know what a telescope looks like and how it works, but to work with one yourself is something else entirely. It's a long road to travel, to go from detecting radio light to producing an image that you can show to other astronomers, physicists, and the world at large. But it's a special experience when the universe reveals itself.

First the telescope's parabolic dish collects the radio waves from space and focuses them. For our wavelength, the entirety of the dish surface has to be calibrated so that it's accurate to within less than 40 micrometers—here it's even better than that. Via the subreflector, which hangs in front of the telescope on four struts, the waves pass back into the focus cabin behind the dish. There they are channeled through a metallic feed horn into the waveguide of the receiver. The feed horn essentially takes on the function of the horn on an old gramophone. In the receiver the high-frequency signals are mixed down to a lower frequency and then fed into a cable. Through this process, free-floating radio signals become electric waves in a copper wire.

The next step is to store the waves. These days even light can be stored digitally—what an amazing process! First the waves must be filtered again and their frequency made to match the much lower frequency of our equipment. A device from Dan Werthimer's SETI program first built to search for radio signals from extraterrestrials converts the repeatedly filtered radio waves into bits and bytes. The light from the depths of space now corresponds to a pixelated sequence of virtual towers that are either zero, one, two, or three blocks tall. It's true the height of the towers is only a very rough approximation of the oscillation of the radio waves, but we have a whole lot of towers and a whole lot of oscillations.

The quantities of data that we record are enormous: 32 gigabits per second—that's 32 billion zeros and ones per second. If we were to draw our data towers with millimeter-thin lines on paper, then after a little more than two seconds we would need a roll of paper that wrapped around the entire globe. Thankfully by now punched paper tape has been replaced by hard drives. The digital revolution has played into the EHT's hands in a big way.

After we've recorded our measurements, the hard drives are simply sent to Boston and Bonn by mail, where the data are further processed. Only after a lengthy process does a tiny image emerge from the giant quantity of data—talk about data reduction! Really we're only recording static: static from the sky, receiver static, and a small bit of static from the edge of the black hole. Thankfully a large part of the sky and receiver static can be filtered out when the data are processed afterward. The total energy of the static that such a telescope gathers from our cosmic radio source in one night is incomprehensibly small. It's the equivalent of the energy produced by a strand of hair one millimeter long that falls from a height of half a millimeter in a vacuum onto a glass plate. The impact will hardly scratch the glass, but we can measure it.

In order for the data to be precisely combined afterward, every telescope needs an absolutely precise clock, and precise clocks are built in Switzerland, naturally—that's true both in general and in the realm of physics. In our case, what we're talking about here isn't a mechanical masterpiece, but rather a highly accurate chronometer from the age of quantum physics. Neuchâtel, near Bern, is one of the main production centers for such instruments. It's where they built many of the atomic clocks that were used in the Galileo satellites, the European alternative to GPS. We, too, had made-in-Neuchâtel atomic clocks: hydrogen maser clocks with a per-unit price in the five figures.

If you're an astronomer who wants to work with a telescope, there's one person you shouldn't mess with, and that's the operator. As the name suggests, he or she operates the telescope, standing at the tiller like a captain on a ship. They know their telescope in and out and steer the dish from their control room, where they sit in front of a wall of screens. At the SMT there are always two operators on the mountain at the same time, who alternated twelve-hour shifts. They are from the area and are accustomed to the solitary life on Mount Graham, whose name in the Apache language means "Big Seated Mountain."

For certain measurements, operators can also hand the virtual tiller over to the astronomers in the room, but they'll take over again immediately if problems arise or a strong wind prevents further operation.

For VLBI experiments there is a strict itinerary for every telescope that we have to stick to. In theory it's automated, since, after all, the telescopes are all supposed to be pointed at the same radio source at the same time, with split-second precision. To prevent any confusion around time zones, all the times are given in Universal Time—the time zone in which the Royal Greenwich Observatory in England, which has long since become a museum, is located.

When taking our measurements, we don't just observe the Galactic Center or the heart of the M87 Galaxy. In between sessions, the radio dishes swing again and again toward calibration sources in order to set the measuring sensitivity of our telescopes. Often we use well-known quasars or galaxies for this purpose. One of these, for example, is the galaxy known as 3C 84, located 240 million light-years from the Milky Way in the Perseus cluster. Herschel discovered it back in the late eighteenth century. The 3C 84 Galaxy is a reliable and strong radio source.

Often bearings are taken on three or four different quasars over the course of a session. Only in this way can we calibrate our entire system. Even atomic clocks are too imprecise for VLBI—we correct them with the help of these cosmic sources and thus make sure that, after the fact, all the clocks are ticking in unison.

Changing the direction the dish is pointed in can take a few minutes. To fill the time, the operators in Arizona have thought up a little gag:[19] whenever the telescope is moving, the lively song "Classical Gas" from the Australian movie *The Dish* plays in the control room and in the kitchen. The movie is about the 64-meter dish in Parkes that picked up the first television images of the moon landing. Anyone who has observed stars or black holes on Mount Graham in Arizona won't ever get this earworm out of their head again.

Measurement shifts can be stressful sometimes. Often the telescope or our instruments have to be readjusted. Atmospheric refractions cause the source to shift slightly and appear in a different spot, and fluctuations in temperature cause changes to the massive dish's viewing direction that are minimal but noticeable. These are all sources of error that we absolutely want to identify and avoid. The process of "pointing," along with the focusing of the telescope, must be corrected regularly during the breaks with the help of the bright calibration sources—most often black holes. Sometimes, when the weather is bad, we don't find the source we want right away, or we find it and lose it again. Then, like someone searching with binoculars at twilight, we just have to keep trying until we've got it again.

The position of radio sources in the sky changes constantly—after all, the Earth keeps turning. Our job is to continually follow the wandering path the radio sources take with the telescope. Another issue is that stars and black holes rise and set earlier in the sky over Spain than they do in Arizona. For a VLBI experiment with several telescopes around the world this poses difficulties, because from their respective positions the telescopes can't all observe the same object at exactly the same time. Sometimes the joint observation sessions only overlap for short periods.

Every now and again the telescope itself won't cooperate. Telescopes are only human, I always say. On March 21, 2015, we report: "The weather looks good." And so we can begin our observations on time. Barely an hour later we suddenly start having technical problems. "The telescope isn't working right. The operators have to deviate from the itinerary in order to repair it," we set down in the log.

Another time the telescope has to stop because the cables aren't long enough for an additional rotation. As a rule, telescopes are designed to be capable of completing one and a half rotations. While following an object in the sky, they can only turn in the same direction for so long. When the maximum is reached, the operator has to spin the whole facility around again to unravel all the cables. Then the ditty from *The Dish* plays for minutes and minutes. I wait, annoyed, for us to be able to start again; we have to miss at least one sequence of measurements and go right to the next item on the schedule.

At the end of this week, I leave Mount Graham with mixed feelings. A lot worked, we learned a few things, the weather was so-so. Tired but satisfied, I drive back down the mountain. Months later we find out that a few components weren't optimally fitted yet, and the quality of the data isn't good.

In spring 2016 comes the second general test. In the intervening time some of the telescopes have gotten technical upgrades. But what is especially important is the fact that we will be integrating the ALMA Telescope in Chile into our network for the first time, on a trial basis. If we can demonstrate that everything went well with ALMA this time, we can go ahead with a measurement attempt in 2017 that won't be a general test but rather the actual premiere for our big project.

Early that year, before we can start, a scientific bomb drops. The

LIGO/Virgo Collaboration announces a press conference for February 11, 2016. We expect a sensation. The secret has already leaked and begun circulating among those in the field, but still we stand transfixed in front of large screens in our university's auditorium and watch the remarkable announcement together with many others around the world.[20] For the first time, scientists have succeeded in directly observing gravitational waves produced by the merger of two black holes. An incredibly weak tremor in space was detected here on Earth. The black holes that merged were 30 times heavier than our sun, though still 200,000 times lighter than the black hole at the center of the Milky Way. "For the first time we've been able to 'hear' black holes," I say excitedly. "Now we want to see one!"

What amazes me is what incredible luck my colleagues had. Before they took their measurements no one was sure that melting black holes of this size could even exist. The gravitational waves' signal was much stronger than expected, and moreover the discovery was made at the end of a test run. If the scientists had stopped it just a few hours sooner, they wouldn't have picked up any of the necessary data.[21] Afterward they never found that strong a signal again. "We'll never get that lucky," I think enviously. "When we conduct our big experiment next year the weather is guaranteed to be bad, the telescopes will break, and the black hole in M87 will turn out to be much smaller than we thought." I brace myself for a long trial of my patience.

Two months later I drive back up the narrow road to the Submillimeter Telescope in Arizona. This time my grad students Michael Janßen and Sara Issaoun[22] are also part of our team. Michael comes from the idyllic town of Kalkar on the Lower Rhine and has written an excellent master's thesis as my advisee. Sara comes from a family of Algerian Berbers. Her parents are both engineers and immigrated to Quebec during a period of unrest in Algeria when she was little.

Later her parents moved again, to Arnhem, where they work for a nearby high-tech firm.

Sara began studying physics at McGill University in Montreal and came to me over the semester break to ask whether I had any work for her. I gave her the data from the Arizona measurement campaign in 2015 and was surprised at what she made of them. In no time she had improved the calibration curves—or completely redrawn them. She even found errors in the telescope software. I realized that Sara was an uncommonly gifted astrophysicist. In 2016, I take her along to the telescope in Arizona and after three days I am almost out of a job: Michael has automated the software operation, while Sara, the youngest member of the team, has all but completely taken over control of the telescope. Meanwhile she is still improving my old calibration measurements from 2015. In the coming years, control of the telescope will remain in her hands. And she hasn't even started her PhD yet. Ultimately the entire EHT will profit from her and Michael's work, as later they will take over the calibration procedures for all the observatories. Both will receive a special commendation from the collaboration.

Technically the results of the test measurements are a success and lead to us getting the long-sought-after okay from ALMA, but they are never analyzed in detail or published. The two measurement campaigns have tested the mettle of our team and the EHT as a whole. If we want to be successful, this mixed bunch of scientists and technicians from every country and every part of the world will ultimately have to learn to work together. And now we know: if everything falls into place, then in theory we could pull it off next year—with a lot of luck.

STRIKING OUT ON EXPEDITION

THE BIG EXPERIMENT

Now the countdown to our big experiment began. The seed had been planted; the seedling had grown into a healthy plant. The time had come to reap the harvest.

Everyone involved in the EHT waited feverishly for April 2017. After years of preparations, of scientific and political tensions, of solving technical problems, our dream was now within reach. In early April 2017, eight EHT observatories[1] were all to be pointed at the same target in the sky: two of the telescopes were in Chile, two in Hawaii, one each in Spain, Mexico, and Arizona, and one last telescope at the south pole. In addition, Sera Markoff, along with a large group of astronomers, had put together a whole fleet of other telescopes on the ground and in space that would be observing parallel to us. From near-infrared to gamma-ray telescopes, everything was in place. We were set to cover the entire spectrum of light so that we wouldn't miss any type of radio emission.

It would be an expedition of extremes. In Chile, astronomers would have to grapple with the dry and thin air at over 5,000 meters above sea level, while those stationed in Antarctica would have to defy the icy temperatures—the average yearly temperature there is almost −58 degrees Fahrenheit. Our planned venture into space was a bit reminiscent of the old days of astronomy, when scientists traveled around the world to study the heavens from the best possible observation point and come a bit closer to understanding the secrets of the cosmos. Then, as today, a thin margin separated failure and triumph. Like Guillaume Le Gentil, who chased the Venus transit in India for years without success, we could easily come up empty in our quest to produce an image of a black hole. If the weather and the technology didn't cooperate, we would fail.

It's spring, and many of my colleagues are in the final planning phases. Equipment is sent off, countless emails are written, video conferences are held, and a preliminary test is organized. We start a group chat to make it easier for us to communicate with one another. More and more members join. On March 5, the German astronomer Daniel Michalik sends a message from the South Pole Telescope, where he is spending several winter months in the Antarctic ice. Normally Michalik works for the ESA—before this he took part in the Gaia program—but during these next few months he and a colleague will be conducting EHT measurements. They live so close to the south pole that they can see the pole markers from their kitchen window. The connection down there is very bad, and the first sign of life from the two of them is a Pink Floyd quote, from the song "Comfortably Numb": "Hello? Is there anybody out there?"[2] We're happy to have someone there holding down the fort for us.

Most of my colleagues make their way to their telescopes toward the end of March. Over several days, scientists around the world arrive at their stations for the big moment.[3]

In these days it becomes clear that, as if by some miracle, the EHT has become a truly international, truly cooperative effort. Every time a new face pops up in our chat group, the person is greeted by his or her colleagues. Joy mixes with excitement; the tension rises. But everyone has their own specific tasks to concentrate on—otherwise it wouldn't work.

At no point was the idea of international understanding so clearly visible as it was during the measurement campaign in 2017. Still today, whenever I see photos of the different teams at each telescope, I feel joy at the fact of our having succeeded, almost without even meaning to, in bringing so many different people together. The two pacifists Einstein and Eddington would have been pleased. And indeed, things remained fairly peaceful during the whole campaign. That wasn't always the case beforehand, nor would it be afterward.

On April 3, I board a plane in Düsseldorf bound for Málaga. From there I take the bus to Granada, site of the Spanish branch of IRAM. The city is just getting ready for Easter celebrations. There's a special atmosphere in the streets as *La Semana Santa*, Holy Week, approaches. Spring has arrived, the temperatures are in the comfortable double-digit range. I'd like to stay longer in Granada and check out the festivities, but I don't have time to take in impressions of Andalusia. I can already see my destination on the horizon, where the snow-covered peaks of the Sierra Nevada rise up to the southeast. In Granada you can drive to the beach in the morning and go skiing in the mountains in the afternoon.

The IRAM Telescope lies a short ways below the summit of Pico del Veleta. At 3,396 meters, it is the third-tallest mountain in Spain. In the Middle Ages the Moors looked up at its peak from the famous Alhambra palace. The route up the mountain holds fascination for cyclists as the way up to the highest paved road in Europe. The journey to the telescope in Arizona was something special, but to travel to the IRAM 30-meter Telescope in April is something else entirely.

Thomas Krichbaum from Bonn is again part of our small group. Krichbaum has undertaken observations here several times before. We ride up to the mountain together in a van. On the way to the telescope we gain 2,000 meters in elevation over just 30 kilometers. The view of Granada from the road is spectacular. Our driver stops for a few minutes so that we can get used to the air up here—a good opportunity for him to drink a cup of coffee, and for me to enjoy the majestic view.

On the mountain up above we can already see the imposing telescope. After being dropped off at the ski lift, we trudge through the snow past a ski school to a red piste caterpillar with white IRAM insignia on the sides. We put our suitcases in a black metal basket and climb into the cabin. The caterpillar takes us the last few meters through the snow up to the telescope in radiant sunshine. The last bit of slushy snow lies on the slopes; a blue sky opens up above us. This is all very promising.

It's always impressive to see research stations built with tons of concrete and steel in the most remote places on Earth. What pains people will go to in order to explore and broaden their horizons! The IRAM Telescope is a monument to curiosity, towering at 2,920 meters above sea level.

The telescope was built here in the '70s, when the first ski lifts were springing up, at the urging of the director of the Max Planck Institute at the time.[4] The director was a passionate skier, and there were rumors that this might have influenced the choice of location.

Today the mountain is covered in skiers and ski lifts. But no building is as large as this one. The facility is far bigger than the rotating building with detachable walls in Arizona. The snow-white structure of the IRAM Telescope looks like you imagine a classic radio telescope to look. The dish, 30 meters in diameter, sits atop a cone-shaped building that can be entered through a large garage door. Because it is located outdoors, the reflector is completely heatable, to keep it free from snow and ice.

Right next to the telescope stands a three-story concrete building whose windows offer a breathtaking view of the Sierra Nevada. This is where we live and work. Sometimes, when the clouds are low, your gaze sweeps over the cloudscape as if you were in an airplane. Time and time again the whole telescope will vanish in the fog, so that you can't even see its tip from the control room—but when we arrive there's not a trace of cloud or fog in sight.

Five people make up the EHT team on Pico del Veleta.[5] I just miss Helge Rottmann, technology expert from Bonn. Rottmann and I had been grad students together. He had checked on the technology here on Pico before we came and then traveled straight on to APEX in Chile.

Compared to Arizona, the IRAM 30-meter Telescope is a four-star hotel, albeit one with 1970s-style flair. The roomy building boasts much more space than the telescope on Mount Graham, and the stairs are always in the same spot. There's a shared kitchen, and we can help ourselves to the refrigerator and pantry, which are always full. We don't need much, though, because a rotation of catering teams prepare Andalusian dishes for us. The locals in these teams seem to be competing to see who can cook the most delicious food. Right away we're served soup, meatballs with grilled peppers, and rich desserts. A giant serrano ham is just sitting there for us to help ourselves to at any time, by slicing off a few paper-thin slices with a long sharp knife. In addition, there are Spanish cheeses and fresh grapes. No wonder that between the technicians, cleaning crews, and astronomers, the mood is good across the board. If you don't watch out, you can put on an enormous amount of weight up here and feel like the inflated chips bag on Mount Graham in Arizona.

There's even a little mascot. A fox who stalks through the mountains nearby caught wind of the good culinary supply situation. Apparently one day he actually managed to steal the serrano ham, or so I'm told. Although there's a strict prohibition on feeding, he still comes by regularly. Is everyone keeping to it?

There's no Wi-Fi up here because the radiation could influence the sensitive electronics. That's a problem for phone junkies like myself.

Often there are a good dozen people on the mountain to operate the telescope. In addition to us, there are two other research teams with whom we share the observing time. An hour of observing costs about 500 euros. Ideally, the facility should be in use twenty-four hours a day. If the weather is too bad for us—here or in another part of the world—then another group uses the telescope.

Beginning on April 4, 2017, we have a window of ten days available to us to carry out our measurements. Vincent Fish, together with Thomas Krichbaum, has again put together an observation program for each day, each telescope, and each radio source—laborious, painstaking work. During this window of time, the weather at all the telescopes around the entire world has to be good enough for us to carry out our program for at least five of the ten nights. Experience teaches us that this rarely happens. To help us coordinate better, my colleague Daan van Rossum in Nijmegen has programmed an online network for the EHT. Here the data from all the telescopes flow together so that everyone can see the conditions at the other telescopes. On this platform, Gertie Geertsema, a trained astrophysicist working for the Dutch weather service, makes available to us the forecasts for all the telescopes from the European Centre for Medium-Range Weather Forecasts' global weather model. A first glance at the forecast shows something amazing: in the next three days there's supposed to be excellent, cloudless weather almost everywhere. I'm a bit worried, though, that our weather information might not be correct. Can the Dutch really predict the weather in the mountains?

Shep Doeleman has set up an analog communications center in Boston; all the telescope information should flow together there as well. A film team is also standing by to record material for a documentary.[6]

Whether we're actually going to conduct our observations or not will be decided at a teleconference that we hold four hours before the start of each planned observation session. In the past few years the whole thing always went somewhat chaotically. How would it go this time? Coordinating eight telescopes and dozens of academics is about as easy as taking an elementary school class of spoiled, neurotic city kids through a candy store during Lent. Neither group does as they're told.

On April 4, all the EHT telescopes conduct a test run. Within a VLBI network there are always technical failures. We have to turn the telescopes into team players. Just like anything to do with computers, though, it's usually the user, that is, the human, who is the weakest link in the chain. The tiniest errors stand in the way of success: two cables get mixed up, data are labeled incorrectly, the wrong command is given unwittingly. We're all tense. The technicians check their instruments again. "Fringes between JCMT and SMA," Remo Tilanus writes excitedly to our group chat from Hawaii. He's ended this sentence with four exclamation points. "Some dumb, obscure box had to be turned off and then on again," he adds. Finding "fringes" means that at least these two telescopes, the James Clerk Maxwell Telescope (JCMT) and the Submillimeter Array (SMA) on Mauna Kea in Hawaii, are working well together. Things look promising for the following day. I go to bed to recharge my batteries. Looking at the weather forecast, I know that tomorrow things get serious. Thomas

Krichbaum is still pacing around the control room. "Aren't you going to go to bed?" I ask him. "Oh, I'm just making sure everything's in order," he answers absently.

The tension increases the next day. All over the world, there's always some little problem with the Mark 6 data recorders, which like to act up. Experts in Boston try to help. Everything on our end seems to be in order. Krichbaum writes the group: "The Mark 6s on Pico are booted up and running." At dinner the decisive teleconference takes place. The miracle has happened. The weather is excellent for every telescope around the world—and the technology is working!

"GO for VLBI, this is not a test," Doeleman writes all of us. The measurement session is to begin at precisely 22:31 UTC—in Spain that's a half hour past midnight. Because we're located farthest to the east, where not only the sun but also black holes rise first, we're part of the first group of telescopes to begin. The clock is ticking.

Naturally our entire crew is in the telescope control room long before start time. No one wants to miss this moment. Here we have the poison-green-and-silver control panel that steers the 30-meter dish. It's fitted out with lots of big knobs and switches and seems like it was dropped in here from another era—or maybe from a 1970s James Bond movie. Two clocks show local time and stellar time; a big red button brings the facility to a complete stop in an emergency. Only the four computer screens on the panel give any indication that the technology is now on a whole other level.

My colleagues start to get nervous, and it's no different for me. Everything should—no—everything *has* to work now. We check the instruments again and again. Are our hard drives programmed right, and do they know at what time they should start recording? Is the observation plan ready, and is it really the right one? Is the telescope's receiver sending data? Is it properly focused? I look constantly at the numbers showing water vapor levels in the atmosphere that are displayed in the control room. I can hardly believe how good they are. To make sure, and to calm my nerves, I keep going outside and looking up at the clear starry sky. Not a cloud in sight.

Krichbaum takes charge and sits down in front of the screens in the observer's seat. He takes advantage of the time we have left and starts carrying out the first calibration measurements. I sit down next to him. It's just like it was back when I was a young student, sitting next to him as he explained how to operate a radio telescope. Some things don't change, not even after twenty-five years. I sit back and enjoy it. He steers the telescope's field of view across a cosmic radio source. We can see clearly how the radio emission first picks up and then subsides—relief. The signal is strong, as is our sense of eager expectation.

The first radio source on our list is OJ 287, a quasar 3.5 billion light-years away in the Cancer constellation that contains one of the largest known black holes. There are some who think it's actually two black holes that orbit each other.[7] OJ 287 is our warm-up, so to speak, and we can see its radio emission clearly and distinctly on the screen. Actually what we see on the screen is just two bell curves—that's all. The first "VLBI scan," as we call our measurements, is supposed to last seven minutes. It's 12:31 a.m.; now it's finally time to start. The display on the screens changes and signals that the telescope has automatically switched over to VLBI mode.

I jog over to the adjoining machine room. There the Mark 6 recorders are parked on a man-high instrument rack and making an audible noise, fans whirring, lights flashing. The green bulbs on the front blink nervously in rapid intervals—the data are flowing; the recorders have sprung into action. In the anteroom is the display of the analog-to-digital converter that converts the radio waves into zeros and ones, and the display of the atomic clock. Everything is in order here as well. I'm relieved.

Helge Rottmann logs into the system from Chile. "All four recorders on Pico are recording," he writes our group chat. We know it already. For the rest of the night I'll keep running back in here to check. What's saved on these hard drives will be the difference between success and failure. But we won't know what exactly they're capturing for many months. That's how long the analysis will take.

Now the monotony of observation sets in. Krichbaum follows every movement the telescope makes and takes note of it by hand in his log book. Thomas is an observer of the old school. He's in his element sitting in a telescope's control room. "You still write everything down by hand?" I ask him, amazed. He'd done it way back in Effelsberg, and asked that I do it as well—though I'd secretly written a program that did it for me automatically. "Oh, it keeps you awake. At four in the morning you can make stupid mistakes; it's better to stay busy," he answers. With a sigh, I pick up a pencil.

For observations in the EHT network, there's no point at which all eight telescopes are conducting measurements simultaneously. The reason is that it's not possible to observe the radio sources on our scan list at the same time from every location. As a result, the beginning of tonight's measurement program seems notably Spanish. ALMA and APEX in Chile, the Large Millimeter Telescope (LMT) in Mexico, the SMT in Arizona—all in former Spanish colonies—and us here at the IRAM Telescope in Spain: at 22:31 UTC on the dot we're all staring at the very same quasar out in space. The two stations in Hawaii and the telescope at the south pole still have some time to wait.

Slowly I start to calm down. We're on the right track. Now the question is, how are things going at the other stations? The messages that trickle one by one into our chat give reason for hope. "Everything's going well at APEX." Our colleagues in Mexico, who have started their shift by now, write: "LMT recording, scan checks OK." Doeleman reports from Boston that everything's looking normal in Arizona as well. But what about the crucial ALMA Telescope? For a long time there's radio silence out of Chile. But then the reassuring words come in: "ALMA has observed all scans so far."

Minor problems keep popping up, but that's normal. Mexico reports difficulties in observing OJ 287. They can't focus properly. Same as with a camera, our colleagues don't have enough sharpness on their dish. The telescope network as a whole is large enough to compensate for this, but afterward the calibration team will have their work cut out for them in finding a way to adequately compensate for this loss of signal. Next though, the team in Mexico keeps having to shut down—for whatever reason the motors aren't getting enough power. They have to wait and start over again from the beginning. We lose their data for now, but we've still got all night.

Meanwhile, back in our control room, we're now approaching the key phase. The plan calls for the first observation of M87. It's quarter to three in the morning. Now we have to get our telescope to sniff out the radio emission originating from the precise center of a giant galaxy 55 million light-years away, from a spot where a giant active black hole is flexing its strength—and we still don't know exactly how big the black hole really is.

"Source M87" is the command I enter into the computer. The screen shows the following coordinates: right ascension 12h 30m 49.4s; declination +12°23'28". The telescope turns slowly toward the Virgo constellation, the second largest in the sky. We follow the motion on the display, transfixed. The telescope positions itself to find the right azimuth, or angle along the horizon; its elevation, or vertical angle, is perfect now as well. Same as with any other large movement, we have to do a little adjusting afterward. "Pico Veleta on source

M87 and recording, pointing on nearby 3C 273," Krichbaum reports. The plots in the control room show plausible signal levels; the hard drives spin and fill up. A reassuring sign. On our instruments we see how the telescope follows the center of M87, moving counter to the Earth's rotation. For hours now we swivel back and forth between M87 and a calibration quasar for scans of a few minutes each. Now everything seems to take care of itself.

We get to the point where we're all in a kind of trance. It's a strange feeling. Really the weariness is getting the better of us, but at the same time we're intent on following even the slightest movement of the telescope. I'm actually supposed to take over from Krichbaum in the early morning hours so that he can finally get some sleep. That's what we'd agreed on anyway. But he has a hard time pulling himself away. He stays at his observation post all night; his body must have a secret reserve of energy stored up for such situations.

Morning dawns. At 6:50 a.m., the telescopes in Hawaii awaken and we observe M87 together for a short while. An incredible 10,907 kilometers separate us—the longest distance in the network for this source! We're now pointed just above the mountains—will that be an issue? Fifteen more minutes and the last of a total of thirty-four scans is on our recorders. The lights stop blinking. Krichbaum sends off the last report to the group chat: "All VLBI scans completed. Now we're a bit tired and are going to take a break after thirty-eight hours of observation." The poor guy didn't sleep the night before either. While we drag ourselves to bed, Sara Issaoun in Arizona and her colleagues in Hawaii keep going for hours more. Arizona, as the most central telescope, has the longest shift, and just like us, Sara can't tear herself away from the screen. She'll observe for a good nineteen hours straight—plus preparation time. She's still working when we get up to prepare for the next night. Her break will have to be a short one.

FINAL COUNTDOWN

The first round was a success. If only we can keep this up! In Boston a pleased Shep Doeleman compliments us on a very good first phase of observations and wishes us all good night for now—though most of us will only be able to get a few hours of sleep during the day.

While the others go to bed, a completely new set of tasks awaits me. Shortly before our expedition began, an article about us appeared unexpectedly on the BBC's website, claiming that we were "on the verge" of obtaining the first image of a black hole. How did they know this? We hadn't sent out any press release; we have no idea if what we're doing is even going to work. What if we just end up with another fuzzy splotch? Aren't we raising expectations too high? What if we have to take measurements over many years? I'm not media-shy, but having this article come out now is one distraction too many.

Nevertheless, the article is enough to trigger a real media avalanche, and we send out a press release after the fact. Now my phone is ringing off the hook, with journalists all over the world wanting to know how things are going for us on Pico del Veleta. I give live interviews over Skype for Sky News and Al Jazeera and speak with Dutch radio. Our social media pages start to light up as well, so much so that old-guard radio astronomers start to grumble about it on these very sites. I try to temper expectations. We'll be satisfied if we see something that resembles an ugly peanut, I say.

There's definitely no chance of getting enough sleep now, which takes its toll on my language skills. A few Dutch Twitter users take issue with my grammar. Only when they hear that I'm actually German do they seem a bit more forgiving. I almost take it as a compliment.

That evening the next decision on the weather approaches. Everywhere the conditions for observing are exemplary. After a short teleconference, the start signal for the second round comes as expected: once more it's "a go for VLBI." Almost a hundred scans are on the docket tonight, meaning we've got an even more strenuous shift ahead of us. We start in the dead of night, two hours later than the day before. For hours we swivel back and forth every few minutes between 3C 273 and M87. Thankfully the weather holds. We finish the last scan of the night right on schedule at 7:30 in the morning. At this time the center of our galaxy M87 is at an angle of just 10 degrees above the horizon. Even the strongest radio source sets eventually.

The observations come with plenty of entertainment. The group chat is lighting up—the longer the night stretches on, the more we start fooling around. Everyone is clearly in a good mood. Suddenly Daniel Michalik sends a photo from the south pole showing him and his colleague[8] in their thick jackets and ski goggles standing in front of the massive receiver cabin of the South Pole Telescope. Behind them the broad expanse of flat landscape opens up all the way to the horizon: hundreds of kilometers of ice and snow stretching on into nothingness. The white of the landscape passes directly into the blue of the sky. The image has a very particular aesthetic and would deserve a special place in any museum of technology. Now I have a sense of what unbelievable conditions the two of them are working under down there. The temperature is an unimaginable –80 degrees Fahrenheit, writes Michalik.

The pictures are great. I want to see more of them. Right there on the spot I decide to inaugurate an EHT beauty contest. In a separate thread I ask everyone to post photos of their impressions from their telescopes. This way we'll all get a better idea of our colleagues' workplaces—and, in some cases, of our colleagues themselves. Some of us in the EHT hadn't ever met in person before the expedition. In an entirely casual way, the contest to see who can post the best photo brings the team a bit closer together.

The next day it's not certain we'll be able to conduct measurements. The weather near some of the telescopes is touch-and-go, so the decision is postponed for now. The operators in Spain start to get uneasy. If the decision isn't made soon, another group will work with the telescope and we'll lose the night. At the last minute the group decides to go ahead. On the one hand, the fact that we can work in every location for the third night in a row is a huge stroke of luck, but the all-night shifts are taking their toll on us. Doeleman writes to acknowledge this and to say that it's clear to him how great the strain is on the whole team. But we should make use of the days we have. We can always sleep later.

This time things get started for us around six in the morning local time. The Galactic Center is finally on the program. Now I'm especially nervous, but everything goes beautifully. The next day our luck with the weather has run out. It's raining in Mexico. Here on Pico del Veleta there's even snow in the forecast. In Arizona, too, the weather is inconsistent, plus there's a strong wind blowing there. Maybe too strong for a large telescope. We decide to take a two-day break. In other years we might have kept at it, but no one's sad about it now. The teams are completely exhausted.

I use the time to develop a little computer program that we can use to prepare and start future measuring sessions. For me programming is like meditation—a wonderfully calming distraction.

After two nights it's again "a go for VLBI"—the weather has settled down at the critical stations. For us it's fantastic tonight. The air here doesn't get any drier. At the south pole, though, the telescope drops out. The team there has to sit a few scans out, but then the facility is up and running again. One more night after this one and we'll have it in the bag. This night also goes smoothly, with just a few disruptions. Around eight o'clock in the morning Pablo Torne reports from Pico: "We're finished. Aside from two missed scans everything went very well here." We've done it. Our project manager Remo Tilanus checks in from Hawaii, thanks everyone for their work, and wishes us a good trip home. "Or have fun skiing on the mountain"—he knows the telescope.

We lean back. The mood in the control room has a solemn air to it. The lack of sleep has left its mark, but still this moment feels like a triumph, at least for the time being—it will be months before we know whether the data are really usable. Still, we've done our part.

With the end of the last shift, the tension starts to ease off the other telescope teams as well. Now it's time to celebrate. The first *Prost* comes from the south pole. They've got a bottle of scotch on hand for the end of the measuring campaign. How did that get down there? At the LMT in Mexico the mood is also clearly excellent. Gopal Narayanan posts that they've just performed their last scan to the musical accompaniment of Queen's "Bohemian Rhapsody." The next song on the playlist is "The Final Countdown." They're feeling good in Boston, too; they're listening to "Somewhere Over the Rainbow" by Israel Kamakawiwo'ole. But my students in Arizona take first prize. Sara Issaoun writes: "Right now we're listening to a song by Muse. It's called 'Supermassive Black Hole.'" This throws me a bit. I know what a supermassive black hole is, but can someone please tell me who "Muse" is?

THE JOURNEY HOME

Expeditions are some of the most exciting experiences in a scientist's career. But because astronomers travel a lot anyway, I always look forward to the moment when the work is done and I can get back to my family as quickly as possible. My wife always says she has to bring me back down to earth first whenever I get back. That's certainly the case after the measurement campaign in Spain. Working through several nights in a row has sapped the strength of all of us on each of the EHT measurement teams. What I need now is some sleep, and I'm sure it's no different for my colleagues at the other telescopes. While the swarm of EHT researchers make their way back to civilization from the remote edges of the Earth, the hard drives from the data recorders take a journey of their own. Nearly a thousand of them are sent off in large wooden crates via courier service.

"Please don't let anything go wrong," I think. It wouldn't be the first time that VLBI data had gotten lost in transit. The EHT's entire treasure trove of data is saved on these drives. There is no backup; the quantities of data are too large. We have to work without a net or harness. The loss of the data would be an unimaginable catastrophe for us and our project. The campaign would have been all for nothing, and who knows when we would have such good weather again.

One by one the crates full of hard drives arrive at Haystack Observatory. Some of the data is sent from there to the Max Planck Institute in Bonn. We have to wait longest for the hard drives from the south pole. Half a year has passed before the Antarctic winter has ended and they can be flown from McMurdo Station. Our team members at the south pole also have to wait that long. Their night is far from over.

In five days of observation each of the eight telescopes gathered about 450 terabytes of data. That means we have to analyze about 3.5 petabytes—a petabyte is a number with 15 zeros in it! The first step is to correlate the data from the different telescopes, meaning to layer and combine them in precise chronological manner. The teams in Massachusetts and Bonn plan to divvy up the work between them, and each team will also double-check a portion of the other team's data. Both institutes have decades of experience doing this kind of work, but still: better safe than sorry.

The point of correlation is to fish out the radio waves stored in the vast sea of data and layer them, one on top of the other. The EHT is an interferometer, and the information it produces always results from combining the waves detected by two telescopes—for our purposes, one telescope on its own is worthless. You can picture the interference like this: you throw two stones in a pond, and in so doing produce circular ripples, or waves, in the water. When the waves overlap, they cancel each other out in certain places, while in others they amplify each other and produce a characteristic pattern. The shorthand term for this pattern among radio astronomers is "fringes"—short for "fringe pattern," basically a pattern of intersecting lines. From the direction and strength of the lines, one can get a very precise reading on the relative size of the two stones and the place where they hit the water. To complicate matters, though, radio astronomers don't conduct their measurements in a calm pond, but rather in a stormy ocean. They first have to layer the peaks and troughs of lots and lots of waves before they can see anything at all, and in order to do this the waves have to be precisely synchronized; otherwise they diverge.

To achieve synchronization, you have to shift the radio waves relative to one another in such a way that they oscillate to the same rhythm, so to speak. The experts at the correlation centers are a bit like DJs who line up the beats of two different songs by adjusting the speed of two turntables. The beats are then perfectly synchronized, so that the songs sound like a single track. The difference here is that radio astronomers don't work with two turntables, but rather with the complicated Mark 6 data recorders from a number of different telescopes. Only when these are perfectly in sync do the telescopes start functioning in VLBI mode.

For us that means waiting, whereas DJs know immediately if their two turntables are synced up. If they mess up and the rhythm of the two songs gets a bit out of sync, they notice right away—or at the very latest when the dancers start covering their ears and rushing off the dance floor on account of the jerky beats. But while we're taking measurements, we can hardly check to see whether everything is running synchronously. Our experience while observing is like that of the early explorers on the high seas: without GPS and without landmarks, they navigated across the ocean's endless expanse in the hopes of finding their big destination. We won't know till the very end whether all our efforts will turn out to have been in vain; like the explorers, we'll only know when we arrive safely. Fringes, that is to say, good overlapping patterns, aren't always found. Until the results come in, we wait and nervously bite our fingernails.

To search for the unifying tempo for every telescope, you have to know each telescope's position relative to the sky as well as the relative arrival time of the radio waves from space—and you have to know it with precision. In order to estimate these values, VLBI experts use a model of the Earth's motion that takes into account its rotation, its imbalance, and the measured pole wandering due to the motion of both the seas and the layers of the atmosphere. A supercomputer that interlinks more than a thousand arithmetic units searches the static of radio waves detected by each pair of telescopes for common oscillations by shifting the waves to line up with each other and picking out the best correlations.

Finding the right values is arduous, and there's a lot of room for error. If you're just a millisecond off, then you have to search through millions of alternatives; a complete data correlation analysis often lasts longer than the observations themselves. For that reason, you begin, on a test basis, with a small data set: two telescopes looking at a very bright quasar.

On April 26, 2017, the first tidings come from the depths of the correlator. Mike Titus, correlation expert at MIT, reports proudly of finding the first fringes between the JCMT in Hawaii and the LMT in Mexico for the quasar OJ 287. A weight is lifted. This is like the cry "Land ho!" Now it continues, bit by bit. The next day Titus reports an even stronger interference pattern between the JCMT, the LMT, and the SMA for the radio source 3C 279. Almost every day another successful report comes in; little by little they come to include the entire network. On May 5, I inform the director of IRAM[9] that the EHT has now found fringes between almost all the telescopes. It's unbelievable—it worked, all of it! Even if we have no idea what the data will show us—they're the best that have ever been collected in an experiment of this kind. We're entering new territory in space.

It will be nine months, however, before all the data—including the data from the south pole—have been correlated. Before we have even a clue what the data will tell us, we've already begun our second observation campaign.

In April 2018, luck isn't on our side. We miss the first three days because the new receiver for one of the telescopes isn't ready in time, and after that the weather won't cooperate. On Pico del Veleta I can no longer see the tip of the telescope—it's lost in the fog. The dishes of the ALMA Telescope in Chile suddenly ice over, and in Arizona and Hawaii the weather is only so-so. Still, the new Greenland Telescope (GLT) operated by our colleagues from Taipei takes part for the first time.

Then comes shocking news: we hear that the team in Mexico has been held at gun point by an armed gang. My doctoral student Michael Janßen is with them. I desperately try to get in touch with him, reproaching myself all the while. What dangers have we asked our young colleagues to face? There hadn't been any incidents there before now, but I'm the one responsible for the project and for my colleagues. Finally I reach Michael on the phone. "A dark pickup truck blocked our way when we were trying to drive to the telescope," he tells me hurriedly. "Six heavily armed men wearing masks surrounded us. We put our hands up. One of them spoke a little English. When I realized that he was nervous, I got even more nervous." Michael sounds amazingly composed as he relates what happened to him, but I can tell how shaken up he is. "I tried to explain that we were astronomers. Then there was a lot of confusion. They told us that they were going to protect us and then drove away. Katie Bouman and Lindy Blackburn were already up at the telescope. Luckily nothing happened to them, and I'm safe now, too," he concludes. "You're coming back," I say, "as soon as it's safe to leave." Shep Doeleman and I have a frantic phone conversation, and then I speak with the director of the telescope.[10] We decide to pull our team out and finish the campaign without this important observatory.

No one knew if it was a failed kidnapping attempt or if the secret police were behind the operation. We didn't want to find out, either. Around this time tensions had started to flare up between organized crime in the state of Puebla and the central government in Mexico. Later, along the winding, low-visibility road up to Sierra Negra, a dormant volcano, more and more ambushes began to occur. As a result, in February 2019 the institution that operates the telescope, Mexico's National Institute of Astrophysics, Optics, and Electronics, made the logical decision to shut down the LMT and the nearby HAWC gamma-ray telescope for the time being.[11]

For this and other reasons—technical difficulties with other telescopes—our next scheduled measurement session in 2019 also couldn't take place. We meant to try it again in April 2020, and I planned to travel to the new IRAM Telescope, NOEMA, on the Plateau de Bure in the French Alps. Then the coronavirus put a stop to our campaign. Two weeks before the campaign came the lockdown— again no measurements. 2017 seems to have been our annus mirabilis. The data from 2017 would have to show us if we'd made good use of this unique opportunity.

AN IMAGE RESOLVES

HOW STATIC BECOMES AN IMAGE

Images of outer space don't just fall from the sky. Quite the contrary, every astronomer knows how much effort and patience are necessary to capture an image of the cosmos—especially when the light waves are stored on hard drives. After gathering the data, we essentially have to assemble a globe-spanning telescope on the computer and figure out what the dish or mirror of this giant telescope would do with real waves.

The mathematical operation that a mirror carries out when it focuses light from space is called a Fourier transform. The operation is named after the French mathematician Jean-Baptiste Joseph Fourier, who introduced the idea in 1822, and today it is used in every conceivable area of our day-to-day lives. Anyone who stores compressed JPEG images or MP3 music files on their computer uses aspects of Fourier transforms. Our ears do as well, transforming oscillations into notes. It turns out our ears are mathematical geniuses, as are simple concave mirrors: they can handle complex mathematical operations automatically, in their sleep—as anyone who has been startled awake in the middle of the night by the beeping of a wrongly set alarm clock has experienced. On the computer, though, we first have to complete the arduous task of programming these transforms ourselves, which means teaching the computer to carry out the operation step by step.

A unique quality of the Fourier transform is the ability to leave out information without losing the total impression of an image or a piece of music. Electronic compression processes take advantage of this every day: you make a Fourier transform of the image or the piece of music, deleting unimportant parts of the data and saving the data left over, and at any time you can use these to transform the image or the sound file back to its previous state. The differences are hardly visible or audible, but the data quantities have become substantially smaller, which means, for example, that more images can be stored on a single memory card.

The same thing happens when there's dust on a camera lens, or when we look into the night sky with a telescope with a scratched-up mirror. We lose information as a result, and the mirror can't completely carry out the Fourier transform. Still, we don't get an image with holes or perforations in which certain stars are missing, but rather one in which every star looks a little less clear. Without our noticing, the disruptions caused by missing information are distributed across the entire image. Every flaw in the mirror influences all the stars in the image equally. With a computer algorithm, however, you can for the most part calculate and remove these flaws, and in so doing polish up the image.

For this reason, a global radio interferometer, which rather than being made of a big reflecting mirror is made up of many small telescopes linked together, doesn't have to be complete. It works even if its telescopes don't cover every inch of the world's surface. This is the equivalent of a scratched-up mirror with lots of holes—in fact, a lot more scratches and holes than mirror. Nevertheless, with a little skill and some valuable mathematical knowledge, a precise image can be reconstructed. This saves a lot of antennas and even more money. Totally covering the Earth's surface with radio telescopes would also be probably something of an imposition to some of our fellow Earthlings.

A good comparison for the Fourier transform of an image is a symphony. The image that you see is like the music you hear. The Fourier transform of the image, then, is like the score of the symphony, and a radio interferometer is a measuring device that records the music and divides it back up into the individual notes of the score.

At any given time within our VLBI network, every combination of two telescopes is measuring exactly one note of the image, which the correlator calculates. The distances between the telescope pairs are the baselines; you can imagine them as being like the different-size strings of a harp that are responsible for different notes. Only here it's the other way around: the strings don't produce notes; rather, they listen for them, and the longer they are, the higher the "image-note" is that they pick up. Returning to the symphony analogy, the short baselines would mainly hear the timpani and the double bass, and the long baselines just the piccolos and the triangle.

If you were to carry out a Fourier transform of the image of a person's head, for example, the low image-notes would just capture the shape of the head, but not the facial details. The high image-notes, on the other hand, would make the distinct contours of the mouth and nose visible, but not the head around them. What's important is how long the virtual strings between the telescopes are from the perspective of the radio source. If you look at the string from an angle, it appears shorter than if you look at it from directly above. As a result of the Earth's rotation, the projected string length and the direction change, and over the course of a night of several hours of observation, the telescope gets tuned.

To get a good image from a VLBI network, the measuring sensitivity of each individual telescope must be precisely calibrated relative to every other telescope, and relative delays between the telescopes must be corrected. This is equivalent to adjusting and evenly polishing a mirror that consists of several segments, or precisely tuning a piano. Our calibration group[1] takes on this task in spring 2018. It makes sure we have the right mix, adjusting the different volume levels in a big piece of music with a lot of instruments; it conducts the "sound check" before our concert can begin. Only then can a harmonious, symphonic image of a black hole emerge from the cacophony of data.

One day in mid-May I'm leaving my office as usual when Sara Issaoun comes up to speak to me: "Have you seen our first calibration plots from Sgr A* and M87? I think you'll find them pretty interesting," she says in a suspiciously calm manner. Sara is always in a cheery mood, but today her eyes have a particular roguish gleam to them. I'm curious and take a look at her screen. Then I do a double take. Stunned, I ask, "Do you all believe what you're seeing there?" "Well, sure, it's preliminary data; of course we'll have to examine it more carefully . . . ," she answers.

What the calibration team is looking at is a curve of faint points. It's the musical scale for M87 and shows the volume of every note we measured from the object, arranged by frequency—like the equalizer in an old-fashioned Hi-Fi rack. Here, the volume decreases steadily as it moves in the direction of the image's high notes, and eventually hits zero. If the image of the black hole were a portrait, we would now know exactly how big the head was. The fewer the high notes, the larger the head. But then the curve suddenly starts to go up again. We also measured a lot of loud high notes. The head has a face, too, and we've captured it! The highest—and most crucial—notes reached us in the very last minutes, when we were observing simultaneously in Spain and Hawaii—truly astounding!

I take a deep breath, relieved and yet nervous at the same time. "This is too good to be true!" Every one of us knows the shape of this curve; it's in every radio astronomy textbook.[2] "I don't want to say it, but this corresponds rather exactly to the Fourier transform of a ring. If that's true, then M87 really is as big as some people say it is, and we can see the shadow," I say, almost in awe. "Yes, six to seven billion solar masses," Sara adds, smiling.

"Well, okay, first we'll just have to wait and see," I reply, determined to sound casual and try to put my poker face on. Nevertheless I spend the rest of the day pacing nervously around my office. It's as if you'd received news that a very special guest, whose visit you'd spent decades waiting for, was about to arrive—that's what the situation feels like for me. Soon we would be able to see this guest for the first time. A prayer of thanks fills the otherwise sober room.

THE BIG SURPRISE

The fact that we don't measure every note with VLBI means that in theory there are many images that would be compatible with our measurements. If you don't have all the notes of a symphony, you can theoretically whistle a lot of melodies along to it—though most of them would probably sound very off-key.

I'm worried about how we're going to make sure that we're not deluding ourselves. We have to be our own fiercest critics. Thankfully everyone on the team seems to be distinctly aware of this danger, and we carry out every single step of the analysis with at least two independent methods.

The calibration team works unbelievably hard to process the data. Lindy Blackburn at Harvard, an expert in this kind of job, writes one data pipeline, and Michael Janßen writes a second in collaboration with our team. Michael names his rPICARD.[3] Like my favorite *Star Trek* captain, all you have to do is say "Make it so!" and the data processing proceeds fully automatically. Both pipelines deliver compatible results; now the instrument is tuned. We can now release the data and produce images from them. A large, very committed imaging team[4] pulled from among the entire EHT collaboration now turns its attention to this task.

The path to a scientifically polished image is still a long one. Dozens of colleagues from around the world are involved in the imaging group, and their task entails many different steps. There are numerous possible methods for generating an image, and here Katie Bouman enters the fray. She is an expert in electronic image processing, a subject she was passionate about even as a high school student. After completing her studies, Bouman first worked at MIT before moving to Harvard. She knows all about the ambiguities of image processing and how to safely avoid the biggest pitfalls. For the EHT she organized regular competitions to test out VLBI experts and algorithms. The specialists received simulated data from her. Some looked like you might imagine a black hole to look, others showed a jet, and some resembled a snowman with a hat, scarf, and carrot nose. The teams had to hand in their reconstructed image not knowing what was hidden behind the data—really it was like a miniature beauty contest. There were even juries to judge the teams' work; I sat on one myself on one occasion. In this way we subjected ourselves, again and again, to a kind of quality control for pure data analysis, and the imaging team was able to select a few tried-and-true algorithms, which it then developed further.

Up to now the imaging team has only ever worked with simulated image data and with data from the calibrators. But now it's serious; now the measured, tuned-up notes from the M87 and Sgr A* symphonies are released into their hands. The tension is immense— what would our black holes look like? We feel like kids on Christmas morning. A few giant presents are sitting under the Christmas tree, and now we get to unwrap them. But you can only unwrap this kind of present once. There will only be one first time. In science, unwrapping, or evaluating data, is also an experiment; the fact that it is done by humans can influence the analysis.

The group thus divides into four subgroups in order to unwrap the present independently of one another.[5] I'm on Team II along with my doctoral students Sara, Michael, and Freek Roelofs; the team spans three continents and is led by Sara and our Japanese colleague Kazunori Akiyama.

To ensure that each team comes to a truly independent result, all communication between the subgroups is prohibited. And naturally no one is allowed to show the images produced by his or her group to anyone outside the EHT. We want to be absolutely certain that there are no leaks. I have to confess though, I do still show the images to my wife.

The imaging teams' work proceeds according to a very tight schedule. On the night of June 6, 2018, the measurement data for M87 and Sgr A* are released to the four teams. We're all excited. The doctoral students immediately begin analyzing the data. At first each of them works on his or her own image. At this time I'm back in the US at an AAS conference, where I'm giving a talk about our lunar radio antenna. I try not to let my excitement show, and I stay in touch electronically with Freek and the others on the sly. That night the first images of the black hole appear around the world. No one knows who was first, nor does it matter. The wheels keep turning as I'm sitting on a plane headed back to Germany. I'm so tense during this flight back from Denver that it's almost unbearable. In the in-flight entertainment program I discover a TEDx Talk by Katie Bouman. "By the time I land this will be out of date," I think, smiling to myself. When the plane finally comes to a stop on the runway in Frankfurt, I pull my smartphone out of my pocket to view my group's images. The long-awaited guest is about to arrive.

My emotional state is like the climactic scene of a kitschy nineteenth-century romance novel. The image is like a distant love for whom I've waited decades and whose affections I've only ever experienced in the intense and frequent letters we've written each other. In my head I have an exact image of what she looks like, even though I've never seen her. She is the special guest, and now at long last she is coming to visit. The sight of the first image is like the moment when the carriage finally pulls up, the doors open, and I am permitted to gaze upon the face of my beloved for the first time. Mingled with the joy of expectation are my fears and anxieties. Has my imagination led me astray? Was it all just fantasy? Is reality much more coarse and ugly than I'd thought? What if the sight doesn't move me? The carriage stops a short distance away; the door opens.

Trembling slightly, I open the file that Freek has sent me. It's in a special file format used for astronomy.[6] By this time I'm sitting on the German ICE high-speed train, my laptop in front of me. I take a furtive look around. The other passengers aren't paying the slightest attention to me. The window finally opens, and something gray, something unfocused comes into view. I enlarge the image, adjust the contrast, select my favorite fiery color scale, and then I see it: An unclosed ring? A horseshoe? No, it's more like a three-quarter ring. Isn't that beautiful!

I can't tear myself away; I can't get enough of staring at this sight. It's fascinatingly new, but somehow quite familiar, as if we've known each other for a very long time. For an hour I'm walking on air— then the skepticism returns. This was just one fleeting look! What will it look like tomorrow? And if my first impression is confirmed tomorrow, then there's still the arduous journey to establishing a relationship. Will it last? There's still a long way to go till the wedding.

Soon after that, an email comes in from Kazunori Akiyama. He plans a teleconference with our team for the next day. Everyone on Team II is to compare their images with everyone else's. He drubs it into us that we must password-protect the image files before we send them to him. And he, too, is frightfully excited: "Woohoo! I can't sleep tonight," he writes. I'd like to go straight to Nijmegen to talk to my students. But right now I have to go give a TEDx Talk of my own at the university in Aachen. Before rehearsal, I secretly tuck myself away in a storage room. Hidden among food supplies and chairs, I look at the black holes one after the other. Relief! In each of the images a ring is visible. So it's not fantasy after all. I can't give anything away during my talk. The talk is already out of date as I'm giving it, but I give it with gusto.[7]

In late July the all-important imaging workshop is held at Harvard University[8] in Boston. More than fifty colleagues from every facet of the EHT collaboration meet to show their images: first of the calibration sources, then M87. The meeting takes place in the middle of the summer break. I spend the time off with my wife on the Baltic. In the evenings, though, I'm still glued to my phone, waiting for the latest news. I can't unplug completely, not yet. The images from the three other groups did in fact also display rings, which no longer comes as a surprise but is still enormously reassuring. This young love, so ardent and yet so secretive, has finally been revealed to the lovers' families, where it has immediately found acceptance.

At this point those of us in the EHT's science council begin discussions about how we're going to further analyze and publish our image. Over the course of the summer we'd come to realize that it was much more complicated to process and analyze the observational data from Sagittarius A*. As a result we mean to evaluate only the data from the M87 Galaxy for now. "Let's do the easy part first," says my dear colleague and vice-chairman of the science council, Geoff Bower.

The massive monster in M87 is perfectly suited for our image, because even if its glowing plasma flows around the black hole at near light speed, it takes a few days or even weeks for the gas to complete an orbit, on account of its enormous size. For the eight or so hours that we spend trying to snap a picture with our global telescope, the black hole in M87 holds as still as a chubby hibernating bear. The center of Sagittarius A*, on the other hand, is a thousand times smaller than M87; consequently the hot gas revolves a thousand times more frequently in that same period, and its appearance changes. When we try to take its picture it wriggles and jumps around like a fidgety two-year-old at a birthday party. Every picture with a long exposure would be somewhat blurred, and we'd have a lot more trouble trying to generate a clear and distinct image from our measurement data.

After the workshop at Harvard the blind imaging subgroups are dissolved. Now the whole team starts over from the beginning. We now know roughly what radio object M87 looks like; it's time for the computers to calculate the best possible images. We choose three established algorithms to face off against one another.[9] A team develops more simulated VLBI data that are deceptively similar to the real data, but nevertheless are based off different images. Some are rings, some are disks, some are simply just two blobs. The data automatically glide through the algorithms and the imaging team produces thousands of images. In the end they choose the exact parameters that reproduce all the simulated images equally well—including those without a shadow in the middle. If we chose an algorithm that was only good at reconstructing rings, we'd be cheating ourselves.

Only then does the team take the three algorithms with their newly chosen parameters and deploy them on our real measured data for M87. We get three slightly different but distinct images. I'd never have expected we would get such good pictures. They show a glowing red ring with a dark spot in the middle. The color is no coincidence; it was inspired by the theoretical predictions in our old shadow article. A colleague in Arizona[10] has adjusted the color range a bit and improved it. You might not be able to see radio-frequency radiation, but once our image is published, in the public's perception, black holes will have a red glow. Even NASA will start using red in their computer animation of a black hole.[11] When I later tell the story of the red-colored radio waves to Lothar Kosse, a composer of modern Christian praise & worship music, he says, fascinated, "I'm seeing colors that I didn't know existed." This is a very good way of putting it, I think.

We can go public with these images. It's like the day of a secret engagement—starting today, the preparations for the wedding begin.

The analysis teams had already started preparing to examine the results. Now they're running on all cylinders. The theory groups[12] are working around the clock. Using supercomputers, they produce a giant library of simulated black holes in order to be able to compare them with the data. There have never been such extensive and detailed simulations of black holes.

Another team[13] prepares to measure the black hole. How large is it? Can we deduce the mass? The orientation?

What comes together here in a short period of time is breathtaking. Tales can be told of countless acts of heroism, large and small. Everyone gives their utmost, but the adrenaline and the many sleepless nights take their toll. Here and there someone pushes themselves to the limit or beyond, while another gets left behind or suffers under the pressure. The seemingly heroic round-the-clock dedication also has its aggressive side, because it triggers a dangerous frenzy that wears out both heroes and fellow colleagues alike. Alongside intense global cooperation, we discover occasionally archaic, territorial tribalism, one faction going after another in force because their ideas or methods "aren't from here." The relationships in the management team worsen, and the board and science council have our hands full trying to hold the whole team together. Some pour gas onto the fire, others try to put it out, but in the EHT organization as a whole each and every one of us is still committed to reaching our goal and delivering the best possible work.

As project scientist, Dimitrios Psaltis tries to steer this creative storm into more organized channels and develops a publication plan. Should we publish the image quickly with a short article in one of the big science magazines—maybe in *Nature*? That would be misguided. The image is so spectacular and so trailblazing that we shouldn't open ourselves up to the charge of sensationalism. So much has come out of this, and it should all be documented! After many discussions with us and the entire collaboration, Psaltis puts forward a plan for six academic articles that receives the blessing of the science council. We want to properly portray the whole scientific process that was necessary for the EHT: the VLBI technology, the data calibration, the production of the image, the simulations, and the image analysis—an article for every subject, with an overview article to summarize and classify the findings. We end up with 204 pages—almost a whole book about a single image.

In November 2018, the EHT meets in Nijmegen for a large collaborative session—120 scientists come together at Radboud University.[14] Here all the different aspects of the EHT are to be discussed together. It's the first joint meeting since the last observation campaign and since the collaboration agreement came into force in 2017. Katharina Königstein, my assistant at the time, puts a ton of love and energy into preparing this week. The conference takes place at the

Berchmanianum, an old Jesuit monastery that Radboud University had recently taken over and renovated. Here, under the stern gaze of the saints in the old chapel, we will discuss the line of attack for the EHT and our six articles.

On Monday morning I wait outside the doors of the monastery for the buses carrying my colleagues from their nearby hotels. When the doors open I see so many familiar faces that my heart is flooded with warmth. The mood is cheerful and easygoing. "I know you from TV"—this line gets repeated often. Up to this point, many of us have only seen each other in video conferences—although we've spent countless hours together virtually, we've never met in person before now.

A year and a half later the coronavirus crisis would come along, sending many people into quarantine and completely changing the work culture in many companies. Videoconferences day and night were already completely normal for the EHT, but our self-imposed quarantine was interrupted at the collaboration meeting of 2018 for an intensive week together. This experience was enormously important for the dynamic within the EHT. It showed us all what emotional and social factors are lost when the only contact people have with each other comes via screens, cameras, and microphones. The meeting at Nijmegen was like a class reunion where you see people again with whom you still feel remarkably close but from whom you've nevertheless become somewhat estranged.

In the free time the group is buzzing like a beehive—all over the place people are forming spontaneous groups and chatting. The weather cooperates and reminds us of the perfect days we had for our observations—even though it's usually gray and wet in the Netherlands at this time of year. Following an old VLBI tradition, I insist on a small soccer game. I even score a goal—though climbing the stairs the next day is a struggle. For the highlight of the week,

Katharina Königstein organizes the conference dinner at Nijmegen's Saint Stephen's Church. This strikes me as unusual at first, but in the secularized Netherlands, this is a way for a scientific organization to generate important income for the upkeep of the church, and as a group we have an emotional shared experience here. When an opera singer appears and starts singing from the balcony to the accompaniment of organ music, out come the phones to capture it on video—and tissues to dab at tears.

At the plenary sessions the coordinators for each article present their individual plans. I'm in charge of coordinating the main article summarizing the rest, and I ask the group: "What is our narrative? What story do we tell? What statements do we dare to make?"

We definitely see a shadow—just as you would expect from a black hole. We'll never be able to prove black holes, Psaltis stresses. We will only ever be able to claim that our results coincide with the predictions of the general theory of relativity—but they do this quite impressively. Looking at the images from the theory group, an amazing number of the simulations match our image—especially when we run one-on-one simulations of artificial VLBI observations of the model in the computer. This is both a blessing and a curse. The shadow is, as mentioned, a very robust and very visible sign of black holes. But, for example, we can't really say whether our black hole rotates, and if so how rapidly.

All the same, so long as you can penetrate the gleaming radio fog, you see the black spot, and there is an exact correlation between its size and its mass. The ring is light that is bent around the black hole on all sides. It's brighter on the bottom, just as you would expect when the gas is rotating around the black hole at near light speed and moving toward us. According to the theory of relativity, this movement at near light speed focuses and intensifies the light in the forward direction. Because the jet, and with it the rotational axis of the plasma, points up and to the right past us, and the lower part of the gas is moving toward us, this means the ring has to be rotating clockwise.

Our main result, however, is the size of the ring. In astronomical jargon, the diameter is 42 micro arc seconds. Who would have thought that, after all these years of work and the processing of quadrillions of figures in supercomputers, the answer to the question of all questions would be 42?! It all comes down to this number.

To us on Earth, the black hole appears as large as a mustard seed with a hole in it in New York as seen by someone in Nijmegen, or like a hair from a distance of 350 kilometers. Because M87 is about 55 million light-years away, this equates to a diameter of 100 billion kilometers. By making comparisons to our simulations, we can determine the incredible mass of this beast—it is in fact 6.5 billion solar masses. In terms of our solar system, the event horizon of a black hole this size would have a circumference about four times as large as the orbit of Neptune.

Still, at this point in time it remains unclear which "capture" should and will become *the* image. After all, we have images from four different days and three different algorithms—twelve images total. They all look deceptively alike, but they're not exactly the same. A serious debate breaks out in the imaging group. Geoff Bower and I try to moderate. In the end the group decides to simply average the three different algorithmic methods and use the data from the best measuring day in April 2017 to form a single image. The images from the other days and the individual images for each algorithm will also be shown, but not as prominently. A solution worthy of Solomon. Everyone on the team at large can now claim, and rightly so, that their work is represented in this one image.

The last big question is: When do we publish? Shep Doeleman is planning a date in February, around the time of the big press conference at the annual meeting of the American Association for the Advancement of Science in Washington, the biggest scientific conference in the world. To me this seems too rushed. Dan Marrone and I plead for a date in spring or even summer. Good science, and good scientific articles, takes time, and there are only twenty-four hours in a day. The earlier date quickly proves to be untenable, and we agree on a date in April—after the next planned, and later canceled, observation campaign. Doeleman's rallying cry of "aggressive but achievable" becomes a watchword among insiders; even with the extra time, getting everything ready for April will require a concerted effort from all of us. "Aggressive" is putting it mildly.

BIRTHING PAINS

Before the image can finally be released into the world, we still have another phase of intense work ahead of us. *Astrophysical Journal* has promised to devote a special edition to our six studies. Every article has a group of coordinators—usually the coordinators from the corresponding working group—and often an additional group of subcoordinators, who are in charge of individual sections. The writing is done collaboratively, with several authors working simultaneously using an online platform.

Shep Doeleman, Remo Tilanus, and Vincent Fish coordinate the article about our instrument. Geoff Bower, Dimitrios Psaltis, Luciano Rezzolla, and I coordinate and write the overview article. We're the one team without a working group, since our task is to summarize the whole project—in some cases even before the other groups' results are completely ready. Again and again we produce new drafts and hand them to the rest of the group for further commentary. Every single sentence, every single citation is questioned and sometimes argued back and forth. All the writing teams go through a grinding and strength-sapping process. A publication committee[15] oversees the process and selects reviewers from within the EHT to review each article internally before sending it to the journal.

In our article we don't just summarize all the other articles; we also discuss the strengths and weaknesses of our results. Could it be that the ring is only there by chance—like a smoke ring in the air, caused by the jet, that will soon be blown away? Probably not, because in thousands of VLBI observations of jets around black holes such a thing has never been seen, and our structure appears to be stable. Could it be that there's something there that looks similar to a black hole but that in reality is something completely different? Like a giant conglomeration of still-unknown elementary particles—a boson star? Theoretical physicists have formulated many creative, though hardly substantiated, ideas of this kind, and we did simulate such alternative theories.[16] We can't completely rule it out; the twilight zone around the event horizon could be hiding a still-unknown and more complex physics. At present, however, a black hole is the simplest and most plausible explanation, and it explains a whole host of astrophysical phenomena in the universe.

Our real breakthrough is to be the first to have gotten as close as humanly possible to a supermassive black hole. We can now say that, to the best of our knowledge, the dark monsters of mass in galaxies are in fact black holes.

What the quasar pioneers suspected almost fifty years ago we can now see with our own eyes—and soon the whole world will see it, too. A new phase is beginning: after decades of searching for black holes, we can now begin to measure them. The question now is no longer whether or not they exist, but whether or not we understand them correctly. It's now clear: even if black holes were different than we thought, the discrepancies that would result from that difference would have to be rather small; otherwise our image would look different than it does.

The event horizon is no longer an abstract mathematical concept, as it was in the time of Einstein and Schwarzschild. It has become a concrete place, where we can pursue scientific inquiry. Between gravitational waves, pulsars, and the Event Horizon Telescope, we have now assembled a rich set of tools for closely examining the theory of relativity in the extreme reaches of the universe and on different scales. It is, for example, a fundamental prediction of the general theory of relativity that the size of the event horizon and its shadow are proportional to the mass of the black hole. The gravitational waves that were discovered in 2016 essentially originate from this shadow region, though in that case it was the shadow region of small stellar black holes—whose size one can estimate as a result.

Our black hole is 100 million times heavier, but also 100 million times larger than these small stellar black holes—exactly as we expected. Thus, the most basic prediction of Einstein's theory, its so-called scale invariance, is confirmed, and in particularly impressive, precise fashion, to almost eight decimal points.

As we're writing the articles, I also realize that we have a problem with the name for the black hole in the M87 Galaxy. We simply don't have a term for this gravitational marvel. Astronomers have never given any thought to what such an object should be called. Either we have to give the child a name or forever refer laboriously to the "black hole at the center of M87." M87 is, after all, the name of the whole galaxy and not just the black hole.

And so after extensive discussion with our collaborators, we simply add an asterisk to the name, as with Sagittarius A*. To an astronomer, this seems an efficient and logical decision, and the principle can easily be extended to other galaxies. Later on, though, science reporters won't be at all satisfied with it. People need expressive names for things that they have a connection to, and M87* is far from being a cute pet name. We also thought jokingly about calling the black hole Karl or Albert—as a tiny homage to Schwarzschild or Einstein. But would that really have resonated with the majority of people?

Shortly after our articles appear, the University of Hawai'i will send out a press release announcing that a professor of linguistics has given the black hole the name Pōwehi.[17] This is a word from Hawaiian mythology and means, roughly, "embellished dark source of unending creation." This is a wonderful name; Hawaiians are justifiably proud of it, and it can now become part of their culture. But the image was produced using telescopes located all over the world and belongs to all peoples and all languages. Maybe every country should come up with its own nickname for M87*.

When our study is finished, it runs to nine pages of text. But we need almost the same amount of space to list all the collaborators, institutes, universities, sponsors, and radio telescopes that were involved. All 348 authors are listed in alphabetical order. The first is Kazunori Akiyama, the last Lucy Ziurys, a professor from Arizona who built out and maintained the SMT.

In early February we officially send the article to *Astrophysical Journal.* Now all that's left is the so-called peer-review process, which calls for independent experts to check our results. Normally something like this can last weeks or months, but here the reviewers, the "referees," have already been selected beforehand and are waiting and ready to go. This is the last big obstacle. What happens if the reviewers reject our article or find mistakes that we've overlooked? Some referees react with hostility and can make your life hell. After a few days we get the anonymous reviews back. I hectically skim through the text and fall back in my chair, relieved. The reactions are remarkably positive—all our efforts and self-critique have paid off. We just have to make a few small changes. The other articles also get off relatively unscathed.

Now there are just a few weeks left until our press conferences in early April. In America, Shep Doeleman keeps a firm hold of the reins. He wants to organize a big press event in Washington with the NSF. In Europe we're holding regular video conferences with all the key parties to work out arrangements for our own press conference in Brussels. Plans for further events in Tokyo, Shanghai, Taipei, and Santiago quickly emerge. In Rome, Madrid, Moscow, Nijmegen, and many other cities, the press conference in Brussels is to be streamed live and accompanied by local experts. This way the citizens of every country can hear the results in their own language.

For Europe this is new territory. Usually this kind of press conference takes place at one of the large research institutions like the ESO in Garching or CERN in Geneva. Never before has science made it into the political center of Europe. But this image is also a triumph of European cooperation and financing. At this time the bitter negotiations over Brexit are taking place. With this image, we can set an example of the fellowship of all citizens on this diverse continent. Through their financing and their interest, they have played their own part in the success of the project. This is important to me.

On March 20, 2019, the last paper is accepted. We've long had the press conferences planned. Now, more than anything, the focus is on not letting anything leak to the public. With a project as big as this, with so many collaborators involved, this is extraordinarily difficult. Rumors have been circulating for a while that something major is going to be announced on April 10.

When a science reporter hears that six press conferences are going to be held simultaneously around the world, the alarm bells start to go off. I start getting countless inquiries, day after day, all day long.

Rather than trying me directly, a famous reporter from the *New York Times* calls my doctoral student Sara Issaoun instead and tries to coax information out of her under a pretense, but she keeps mum. He gets his information eventually, though—from America. Most journalists are counting on us presenting an image from the center of the Milky Way. More than a few of them will have to frantically rewrite the article they've written in advance.

The day before the press conference I travel to Brussels with Luciano Rezzolla, Monika Mościbrodzka, Anton Zensus, and his colleague Eduardo Ros to unveil the image. Our diverse group represents five different countries and at least six languages.

In the US, Shep Doeleman and three American colleagues are on their way to Washington, among them my colleague from Amsterdam Sera Markoff. Preparations are underway in Tokyo, Taipei, Shanghai, and Santiago as well. It's another kind of global expedition—only this time we're being closely observed ourselves. At many universities, faculty and students will be watching our event live, just as we watched the gravitational waves announcement a few years earlier. For astronomers this is a bit like the public viewing parties soccer fans go to during the World Cup, only with less beer.

The day before, in the afternoon, we plan the press conference with the help of a media expert—as chance would have it, in the very same room where we once had to defend our financing application before the experts of the ERC. The blinds are shut; I practice my introductory remarks and nervously show the officials from the EU the image on the screen for the first time. Soon I'm looking into rapt faces with eyes agleam, and for a few seconds there's an almost awestruck silence even among the hardened professionals here. I receive an early sense of the emotional power of this image.

That evening I head back to my hotel room to think over my remarks again and practice in front of the mirror. I'd like to say the line

"This is the first ever image of a black hole" in four languages. Sara Issaoun translates it into French for me, German and Dutch I can do myself. As I'm working I get a brief visit from my son Nik, who, despite his young age, has already made a successful start as a musician and film composer. He set to add music to our film sequence zooming in on the black hole for the ESO's website[18] and will include scenes filmed on the day of the conference in his first music video.[19]

Around midnight there's trouble on the PR front. A science reporter and an old college friend of mine[20] discovered our strictly confidential press release containing the image on our unsecure website. He could become an internet sensation with the link, but thankfully he alerts us to the issue. A few of my colleagues spend a sleepless night plugging the leak. Who else has found it? Was he the only one? Tensely we wait for the next day and the beginning of the press conference, but if anyone else saw it, they remain silent, and the day becomes a celebration for science.

In 1992, when the Nobel Prize–winner George Smoot presented the first radio image showing our young universe just 380,000 years after the Big Bang, he said, rather solemnly, "If you're religious, it's like looking at [the face of] God." I wanted to offer my own response to this. If the Big Bang is the beginning of space and time, then black holes signify something like the end. And therefore I end my segment of the press conference with the remark, "It feels like looking at the gates of hell." And the whole world looks with us.

PART IV

Beyond the Limits

*A glimpse into the future: the big,
unresolved questions in physics,
humanity's place in the universe,
and the question of God*

12

BEYOND THE POWER OF
OUR IMAGINATION

AN OVERWHELMING RECEPTION

The impact of the image is overwhelming.[1] It seems to hold fascination for everyone. All the leading newspapers and weekly magazines worldwide report on this unique achievement in the history of science and humankind. It's a leading story on TV news broadcasts; social media platforms are lighting up. It's wonderful and frightening, all at the same time. A shared global moment of joy. A reminder of the emotions that the moon landing brought out in July 1969. My daughter has grown up to become a young curate and is currently doing an internship as a trainee pastor at a school. "Everybody in the teachers' lounge has your image on their phone," she writes me proudly.

It's breathtaking to see how many people immediately make this image their own. It is included in countless photos and shared an innumerable amount of times, repurposed for cat photos and funny memes. Google uses it for the Google Doodle on its home page.[2] It hangs on the bulletin board in the news rooms of leading German media organizations, altered to fit the political anecdote of the day. A photo of a delighted Katie Bouman seeing the image for the first time becomes an internet hit, and she an unwilling social media star.[3] A large stock photo agency in China claims to own the copyright of the image and tries to sell it. The outrage on the internet leads to a 27 percent drop in the company's share price—two days after the press conference, the company had lost 125 million euros in value because of our image.[4]

Never before, it seems, has a single scientific image so quickly excited the imagination of so many people. Ultimately though, our success is also their success: no scientist, no person on the team could have done this alone, without the many people who make sure that we can also get along and do our work: the bakers who bake our bread, the janitors who keep our rooms clean, the catering teams and the technicians at the telescope stations. In the end, every citizen has contributed to this world-spanning project with the support they bring to our global commonwealth.

Later on, many of my colleagues from around the world will tell me in detail how they experienced this incredible day. They all had to explain to friends, neighbors, and the press what it actually was that had come together here. The week after the big press conference passes in a blur of adrenaline. We're all swept up in the frenzy: interviews, talks, and, in between, countless emails and texts. We had all struggled and drawn on the last of our strength to make it to the date of the image's unveiling, and now I'm running on fumes. For the first time in my life I feel a strange tightness in my chest. For weeks now I've felt like a car idling at high RPM, but there's still no break in sight.

I'm scheduled to give seven talks in five days. And this during Holy Week, an especially important time for me, but it doesn't feel like Easter at all yet. On Palm Sunday I give a talk in a packed museum in Nijmegen; on Maundy Thursday I speak to a crowd of astronomers in Cambridge. Again every seat in the room is filled. In the audience is Martin Rees, Britain's Astronomer Royal, who back in the '70s was the first to make the idea of black holes respectable. Now he sees the first image of a black hole with his own eyes and poses the crucial question, "What are we actually seeing in the image? The event horizon?" "Its shadow," I answer, and at that moment I feel like a shadow of my former self. My throat itches, I feel dizzy, I can sense that I'm at the end of my strength. Wearily I drag myself back home—the plan is still to celebrate Easter.

The next day, Good Friday, I go to a service at the Cologne YMCA with my wife, as I do every year. It was in this place that the story of my faith began. We hear the story of the death and suffering of Jesus Christ: welcomed on Palm Sunday by a jubilant crowd, on Maundy Thursday he bids farewell to his friends and is betrayed that same night, and on Good Friday, the innocent victim of a plot against him, he is mocked and crucified. I sit in the last row, listen to the story, and think of the jubilation of the past few days, but also of all the hardship with which it was purchased, the pain of all those involved. I weep. What I need now is the calm and the strength that Easter brings.

It takes a few days for me to get back on my feet. I slowly come back to life, but it's still weeks before the tightness in my chest truly eases.

Without planning to, I give my first big public lecture after Easter at SPRING, a large Christian conference in the Sauerland region. Normally scientific talks at this conference are held in tiny seminar rooms, if that, but the organizers have reserved the large "Hall of Conference" at short notice.

There is no trace of hostility to science in the hall, which is full to bursting; there is only a sense of cordial expectation. Nurses, blue-collar workers, students, retirees, teachers, office workers, and entrepreneurs sit motionless in their chairs, listening intently. My musician friend Lothar Kosse says spiritedly, "Everything is possible; everything is impossible," and slips references to the heavens and a black hole into his next album. Black holes seem to fascinate all people equally. But why is that so?

WHAT BLACK HOLES HAVE TO TELL US ABOUT OURSELVES

Gravity monster, cosmic feeding machine, hellish abyss: no superlative is big enough to describe a black hole. Black holes are the dinosaurs of astrophysics, as popular as the Tyrannosaurus rex, despite or even precisely because of their fearful reputation. The fact that our image of the black hole adorned the covers of magazines the world over was exciting enough. But that people would respond so emotionally to it made things even more moving.

People have told me how they fell under the spell of the image, how they couldn't sleep the night before it was revealed, or how deeply touched and moved they were when they saw it. Neither the Higgs boson nor the discovery of gravitational waves has triggered such an emotional outpouring. So what, then, do black holes have to tell us about ourselves?

It seems to me that they represent fundamental human fears in a way that is unlike any other scientific phenomenon. They are one of the big mysteries in the broad reaches of space. In astrophysics they mark the definitive end; they are quintessential, merciless machines of destruction. People sense this intuitively. In our imagination, black holes symbolize the all-consuming nothingness, a border past which all life and all understanding stops—a glimpse into the hellish abyss indeed.

Black holes tell us of a world that is completely different from our own. A place where light moves not in a straight line but in a circle. If I look ahead, I see my back. For one person, time seems almost to stand still, while for the other it passes normally. Gas spins around at nearly the speed of light and can reach apocalyptic temperatures at which all matter is shattered into single particles. All that is left of molecules and atomic nuclei is a molten hot cloud of protons and electrons—plasma. I could fall into a black hole; I could also, in theory, survive, and even conduct scientific measurements—but I would never be able to tell anyone what I see there. No information leaves a black hole, not even light waves. Black holes bring us closer to the beyond.

And the beyond actually exists—even in physics. In the general theory of relativity, the beyond isn't at all supernatural; rather, it's an important part of the theory that separates the world into two realms: the *here* is the space with which I am connected, from which I can attain information and with which I can communicate. And then there is the *beyond*, the space that lies fundamentally beyond my ability to know it. I learn nothing of it; it stares at me in silence. The two spheres are separated by my horizon.

Black holes stubbornly and fundamentally defy our curiosity and our perceptive faculties. Everything that disappears behind the event horizon stays there for all eternity—assuming at least that Einstein's theory is the last word on the matter.

The eternal beyond that black holes embody poses one of the biggest challenges to modern physics. In theory the space beyond the event horizon has a clearly determined location, and yet it exists only in our imagination. It is just as real as it is completely unreal. Today, with our radio telescopes, we can make visible with astounding precision exactly where in the depths of space the door to the beyond is to be found. We can physically describe it, and even see how light disappears inside and never reappears, forming a dark spot.

"There, it's right there," we can say, "in this precise spot there exists a space that is not of this world." But then we have to drop our hands into our laps, helpless, and admit to ourselves that we can't measure it. Black holes are the beyond in the midst of our here.

For physicists this is akin to declaring bankruptcy. Just what kind of physics prevails in a space that exists in our universe in a clearly defined location but evades any kind of investigation? Is it even natural science? "Yes, of course it's physics," theoretical physicists will say to me, "because we can calculate exactly what transpires in this space!" "No, it's not physics!" I reply. "It's a physics of the beyond!" Or indeed, metaphysics for now.

Most people don't care about the laws of physics when they think of the beyond. But everyone has a notion of what it is, and the knowledge of these notions is ancient. The idea of a world beyond stimulates our imagination, challenges us, and is at the same time closely connected with death. Black holes are just one more symbol in a long series.

More than a 100,000 years ago, our ancestors began to bury their dead. It's possible they already had a notion of a life after death. No one knows what such a notion of the beyond might have looked like. But the first rites of respect and emotion for the dead can be found right there at the beginning, and they bear witness at least to a cultural evolution that led to our current highly developed notions of the beyond.

Many of these beliefs are the same across different cultures: eternal life, divine judgment, heaven and hell. In ancient times people believed in a world above for the living and a world below for the dead, an idea that we see in many different cultures. The Greeks called this place Hades. In Norse mythology the goddess of death, Hel, ruled in the Halls of Hel. This was probably the origin of later ideas of hell. Fallen Vikings, meanwhile, could live in Valhalla, a warriors' paradise.

The Romans also imagined a dark place below; their name for it was Orcus. The Mayans called the underworld Xibalbá—place of fear.

The concept of the beyond became established with the major world religions. Christianity and Islam both teach that there is a paradise, or heaven, and that there is life after death. In Judaism there are two different beliefs: one posits the immortality of the soul, which lives on after death and returns to God. Orthodox Jews on the other hand believe in resurrection. They don't cremate their dead; burial and the repose of the dead are sacred to them. Buddhists and Hindus believe we are reborn millions of times—maybe even as animals or plants. Only by reaching nirvana does the soul break this cycle and bring it to an end.

Black holes add a new, modern myth to all these myths of the beyond. It is a myth inspired by natural science, in which profoundly human questions mix with concepts drawn from modern physics. For us, biological death means crossing a threshold: from the here and now we cross over into a beyond of which we can know nothing; we can't even know if it exists at all. Is there something more to come, or is there just nothing? Anyone who has witnessed the death of a loved one has seen how, in the final minutes of their life, a dying person withdraws from their own body, leaving behind an empty vessel. It is not granted us to learn anything of the final experiences, thoughts, and dreams of this beloved person; they literally take them to the grave and into the beyond with them. "Where is she going?" I asked myself when my mother died before my eyes. A few minutes before, I had still held her hand in mine and prayed with her.

Death shakes us to our core. The trembling before the finality of an ending is one of our most fundamental, one of our most ancient emotions. We try to avoid it, but it continues to exert its almost magical pull on us. Up until now black holes were only ever abstract concepts, figments of the imagination that were brought to life in

Hollywood movies, but now for the first time a black hole has taken on concrete features. We might not be able to touch it or feel it, but we can see it. We can now look the proverbial beast in the eye—and in the same moment, we look into our ancient fear. Is this the first step toward overcoming this fear?

"Look here: this is what it looks like, the entrance to hell," I hear my subconscious whispering. "No reason to panic, you're safe, you're just here sitting in your desk chair." Even if I don't know exactly what transpires there, at least I've seen it with my own eyes—the terror is no longer vague; it can be depicted and described.

As the sun shines through the window and the Earth rotates peacefully on its axis, I look at this image of the black hole and know that it is very far away. No black hole, not even one of the many small ones in our Milky Way, will pull us into the beyond. The closest known black hole is more than four times heavier than the sun and is about a thousand light-years away from the Earth.[5] From a distance its gravitational pull is no greater than that of a normal star, and its event horizon is only about as big as Lake Constance.[6] The chance of us encountering such a small black hole is vanishingly small. It hasn't happened in the last four billion years, and won't happen to us in the near future either.

We can therefore continue to observe black holes from a comfortable distance and take joy in their exotic physics. But a snapshot can develop symbolic power. In that respect our image isn't just science, but also art and myth.[7] The Museum of Modern Art in New York and the Rijksmuseum in Amsterdam both acquired prints of the image for their collection, and a few people here and there have it hanging in their hallway.

Artists are able to capture the abstract in words and images and create reality out of it. Reality, in turn, is changed and interpreted through art. In this regard science also has some aspect of art in it.

The images science makes are never reality itself; they only bear witness to reality and, through the story they tell, create a new, abstract reality, stimulating completely different thoughts and conceptions of the world and prompting new questions.

A scientific image has no value without its story, and our image, too, would be just a dark spot without the story behind it. Therefore the significance of the image lives and dies with the credibility of the people who produce it, and of the story that they give it. Really this applies to every scientific finding. We scientists don't make our living from facts alone, but also from the trust placed in us.

What comes together in the image, then, is the entire history of physics and astronomy's development, plus emotion, mythical excess, intelligent silence, the act of gazing up at the stars, the measurement of the Earth and of space, the understanding of space and time, the most advanced technology, global collaboration, human tension, the fear of being lost, and the hope for something fully new. In every respect this image takes us to our limits—and not all the questions surrounding black holes have been answered, not by a long shot.

The EHT keeps working. We showed that the size of the shadow in M87* didn't change much over a decade, based on a reanalysis of old data with a few telescopes only—as expected for a black hole. We were also able to see how magnetic fields wrap themselves around the black hole in M87*. Everyone is now waiting excitedly to see what the black hole at the center of the Milky Way looks like. Will we be able to take a picture, or will the rapid turbulence of the cosmic soup spoil it for us? Will Sagittarius A* also show us its shadow at some point? What will M87* look like in the next few years? Might we even be able to capture a film, rather than a single image? We want to carry out more observations and are in urgent need of more telescopes. Hopefully we'll soon have one in

Africa[8]—I'm always grateful for any additional support! We will be able to capture definitive and impressively crisp images as soon as we have radio antennas orbiting the Earth.[9] If this happens we really would have a telescope larger than the globe. The world is not enough, there's still a lot to see!

BEYOND EINSTEIN?

WORMHOLES

When I was a little kid I lived with my parents in a large apartment building. In the back courtyard there was a sandbox and a small patch of lawn. The garden was surrounded by an impregnable wall. I always wanted to know what was on the other side, and so at some point I started to bore a hole between the slats with nails and sticks. For my tiny hands, this was tough and arduous work. I worked on it in secret for months, whenever the adults weren't looking. The gap got bigger and bigger, but I never made it all the way through. The wall was simply too strong for me.

When I was big enough to go to school, I was suddenly able to explore the area on the other side of the wall—this was where the school building was. Instead of going through the wall I just had to leave the back courtyard, walk around the block, and then go through the big gate in order to reach what before then had been such mysterious territory. Sometimes you just have to be patient and grow and mature in order to realize that the direct route through the wall is the wrong one, and that the correct route leads around the corner.

I still have the same sense of curiosity when it comes to walls and boundaries. What's on the other side? Will we ever manage to overcome our limitations? Can we go around the dark walls of black holes? Is there not a gap somewhere in the event horizon that we can peek through, or can we possibly take a detour?

Albert Einstein asked the same question in 1935 when he was discussing the interior of black holes with his assistant Nathan Rosen. Mathematically the equations also allow for the opposite of a black hole—a white hole, an object from which things can only come out; nothing can go in. To make things even more complicated, theoretically a white and black hole could be connected to each other by means of a bridge that would make it possible to go into a black hole and come out again through the white hole on the other end.

In physics the construct became known as an "Einstein-Rosen Bridge." In the 1950s, however, Princeton professor John Archibald Wheeler, in a clever bit of marketing, rechristened the hypothetical structures *wormholes*, and in so doing made generations of science-fiction authors happy. According to Einstein and Rosen's construct, it wouldn't just be possible to escape black holes; wormholes would also be able to connect two distant realms of the universe, thus making it possible to travel faster than the speed of light. It would even be conceivable to travel through time and visit another universe.

But is everything that's mathematically possible also real? Mathematics is the mythology of science. It's an abstract method of description that depicts real experiences just as wonderfully as it does fantastic, imaginary creatures. What exists in mathematics can exist in reality—but it doesn't have to. Making the tough distinction between the two—the possible and the actual—is how physicists earn their living.

With regard to both white holes and wormholes, we are confronted with the same question: Mathematically, both appear to be real, but do they also make sense in terms of physics? We have yet to find any indication that a wormhole exists in the universe. While producing the image of M87* we did in fact briefly consider whether it could possibly be a wormhole, but its size didn't fit the predictions.[1]

What makes things even more awkward is that wormholes aren't mathematically stable. If matter flies through them, they collapse—at least according to the theory. In order to prevent this, you'd have to invent a new form of matter that creates antigravity. Antimatter alone wouldn't fit the bill, because it's subject to the same laws of gravity as normal matter. If you toss antimatter into the air, it should fall to the ground[2]—unless it destroys itself beforehand, and matter with it, in a bright annihilating flash.

Another problem is that we have no idea how a wormhole through which it would be possible to pass could form in nature. We'd just have to build one ourselves. For some creative theorists that's not a problem: "Since we know nothing firm about the technologies and materials available to a very advanced civilization, we physicists have an infinity of freedom in building models for traversable wormholes," the Nobel laureate Kip Thorne claimed in the *New York Times*.[3] I'm a bit more skeptical. Even if wormholes could exist, it's not yet guaranteed, even theoretically, that they can really deliver on all their miraculous promises. But we can still dream.

HAWKING RADIATION

Quantum theory and Einstein's theory of relativity are possibly the most groundbreaking theories in all of science. They describe our world in the most essential sense and are both fundamental. To try to unite them, however, is to cross a mental frontier. More than any other celestial object, black holes lay bare this unresolvable conflict.

The general theory of relativity describes the biggest of the big, space-time. Our lives begin and end in space-time; the drama of the whole universe plays out in space-time. It is the theater in which the development of our universe is acted out. In space-time, everything has its own place, its own spot. If we recall the example of the stretched-out bedsheet, however, we remember that this theater does not have a static stage, but rather a flexible backdrop that plays along, that reacts, that changes in response to the actors. The black hole is the most radical actor on this cosmic stage—an actor that literally rips the scenery apart and presents us with profound questions.

Everything has a time and a place? Is that really true? No! Because we still have the second theory, which is just as fundamental: quantum theory. While the theory of relativity describes the very big, quantum theory tells us about the smallest of the small, the composition of matter: molecules, atoms, and elementary particles. Nevertheless, it is the building blocks of light, photons, that make space-time measurable. These light quanta drag space-time out from the darkness of abstract mathematical description into the bright light of perceptible reality. It is here, then, that relativity and quantum physics meet.

But unlike in Einstein's theory, in quantum physics not everything has its own place or its own time. Within brief stretches of time, processes can move forward and backward. Particles can be in two or more places at once, so long as no one's looking. At its most extreme, quantum theory opens up a microscopic world that is just as foreign to us as the macroscopic world on the edge of black holes. Nevertheless, both theories have become components of our daily life and work peacefully alongside each other, for example, in our smartphones. Every chip, every semiconductor in our phones is applied quantum physics. Without quantum physics there would be no internet and no computer processors. At the same time, though, the navigation system on our phones that we use for directions uses results drawn from the general theory of relativity.

But on the edge of black holes, these two theories collide in a fundamental manner. Here a completely new physics must be at work, and for many years tens of thousands of the most brilliant scientists on the planet have been racking their brains trying to figure out what this physics might look like—to this day without clear success.

Up to now the problem has been purely theoretical. Among others, it goes back to the famous astrophysicist Stephen Hawking, who pondered what happens with quantum particles on the event horizon.

Quantum objects are the smallest known rascals in physics. God in his infinite wisdom lets them get away with things that we couldn't even dream of. For example, they can temporarily "borrow" a bit of energy without having to ask anybody's permission. The trick is that they have to give this energy back so quickly that nobody notices.

Essentially you can imagine that empty space is a giant foamy ocean. Again and again, drops of water and sea spray spontaneously shoot up into the air and then drop down to mix back in with the ocean. The border between ocean and air becomes blurred. Near the ocean's surface you get wet even when you're not yet swimming in the water.

In the same way, tiny particles appear in empty space and then vanish again. Empty space is thus not completely empty, but rather full of particle spray. But to create a particle from nothing naturally requires energy. So where do you get it from? On the ocean there's wind, which provides the energy to create water droplets, but there's no wind in empty space. And so nature uses a simple bookkeeping trick: it borrows some energy in the short term in the form of a virtual light quantum. In doing so it produces a quantum pair. This is made up of two exactly opposite twins: a particle and its antiparticle—a miniangel and minidemon, so to speak. The one is positively charged, the other negative. If one spins to the left, the other spins to the right. If one is matter, the other is antimatter. To return to the ocean analogy, the particle is like a small drop of moisture in the air and the antiparticle like an air bubble in the ocean.

If the two meet again, then the properties of each are canceled out, and matter and antimatter destroy each other. Nothing is left over—except for a brief, virtual flash of energy that disappears back into the ocean of space-time. And with that, the energy debt is repaid and nobody can file a complaint.

But it's like a financial crisis. You can keep the scam going only as long as nobody notices and all the debts keep getting paid. When the rough weather hits, things start to go sour. The water droplets are blown over the sea toward the harbor and wash onto the land as spray. The ocean seems to be losing water; people on land get wet. Still, it would take forever for all the water to get blown out of the ocean in this way, and besides, rivers and rain keep refilling it.

This very process also occurs—according to Hawking—on the edge of black holes. In the ocean of space-time, the event horizon is the coast. The storm is the black hole; instead of wind energy there is gravitational energy.

In his public lectures, Hawking described this process roughly like this: sibling pairs of particles and antiparticles are created on the edge of the black hole and borrow energy from its strong gravitational field. Before they can get back together, one of them disappears behind the event horizon. The surviving twin can no longer unite with its antitwin and escapes into the vastness of space. A provisional particle pair has suddenly become a lasting single particle.

But the energy debt that was assumed in order to pay for this particle can no longer be paid back—it's a losing bargain. The black hole has lent out two particles but only gotten one back. As a result it loses energy and mass. It's as if a steady breeze of particle spray is blowing away from it, in the same way you might feel the ocean on the breeze by the coast. It seems as though the black hole were emitting radiation. This is Hawking radiation, which the late British scientist first described in 1975.

Hawking's image of particle and antiparticle is a bit truncated, though; it primarily explains the method of calculation used in quantum theory. Ultimately what is emitted isn't particles, but above all photons—that is, light, and at wavelengths that are larger than the black hole itself. The radiation is also not emitted directly on the event horizon, but rather comes from a broad area surrounding the black hole. Thus it's as if the gravitational field were emitting radiation.

In formal terms you can also describe the radiation emitted by black holes as thermal radiation. A cup of hot coffee will cool off after a certain amount of time, even if you cover it and no water vapor escapes. That's because the cup emits thermal radiation. The atoms on the cup's surface start to vibrate slightly from the heat, and when they do they radiate quantum light particles. The qualities of this radiation were discovered by the German physicist Max Planck in 1900, thus laying the foundation for quantum theory and connecting it to thermodynamics: every nontransparent, black object emits radiation when it's heated up—independent of what it's made of and what shape it is.

A cup of hot coffee is thus quantum physics in action and emits primarily near-infrared light. As a result it loses energy and gradually cools down. Thermographic cameras can see this light; our eyes cannot. But our hand can sense this radiation before we touch the cup. In a manner of speaking, we feel the quantum vibrations inside the cup by way of invisible light.

The mathematical formula for thermal radiation always looks the same and depends only on the temperature: the higher the temperature, the higher the frequency of the light. That's why iron, when heated, first starts to glow invisibly, in the near-infrared spectrum, then visibly red, then yellow, then white—the changing colors represent increasingly higher frequencies. Stars can even shine with a bluish hue because they become even hotter than molten steel.

Theoretically at least, black holes emitting Hawking radiation could be giving off this very same radiation. Thus, you can attribute a temperature to black holes, and this temperature is dependent on their mass. The smaller they are, the hotter they seem. According to Hawking, a black hole with about 0.5 percent of the moon's mass would be about as hot as a freshly brewed cup of black coffee and would also emit the same amount of radiation—though in all likelihood it wouldn't taste the same.

As a result of Hawking radiation, a black hole loses energy, and with it mass—after all, mass is ultimately energy, as Einstein's most famous formula tells us. But unlike the cup of coffee, which cools off as a result of thermal radiation if you just leave it sitting there, a black hole becomes ever hotter as a result of the radiation it emits! The smaller the black hole becomes, the higher its temperature and the more efficiently it emits radiation. At some point it meets its end in an explosion that gives off an almost infinite amount of heat. This could explain why small black holes seem not to exist in nature. A black hole with the mass of two diesel locomotives, a combined 160 tons, would radiate itself into nonexistence within a second.

It's different for astrophysical black holes. A black hole with the mass equivalent of the asteroid Icarus, about 100 million tons, will live about as long as the universe before it expires. A black hole with the mass of the sun would take 10^{67} years, and M87* would need an unimaginable 10^{97} years.

I've tried to find a way to visualize this, but it's just not possible. Let's say you take the mass of the entire known universe—every star, planet, and gas nebula anywhere in space—and gather it all into a giant ocean of matter. Then, once every ten billion years, roughly the age of our universe, you fish out a single, teeny-tiny proton from this ocean of matter, and repeat till it's all gone—the universe would still have disappeared ten million times faster than M87* as a result of Hawking radiation.

What's more, the universe would have to have completely died out, to have become empty and dark, before black holes could completely evaporate, because every gas particle, every light wave in the universe causes a black hole to keep growing. Within spans of time that are much longer than anything we can imagine, supermassive black holes like M87* can only get bigger. The Hawking radiation of M87* is so weak that it wouldn't be possible to build a detector that could find even rudimentary proof of it within the life span of our universe—not even if we could fly there ourselves.

Nevertheless, from a purely theoretical standpoint, black holes could indeed evaporate, and as a result everything that was once trapped inside them would be released. Nothing would be forever—not even black holes.

When calculating Hawking radiation the presence of an event horizon is key, but if Hawking radiation is really a radioactive decay of gravitational fields, I could imagine that ultimately neutron stars or even normal matter might decay in a comparable process, and thus all gravitational fields could eventually dissolve into light. But for now this is still pure speculation.

In the beginning there was light, and at the end light might be all that's left in the universe—unless something new and exciting happens before then.

At the end of the press conference at which we unveiled the image,

EU commissioner Carlos Moedas cited words of Stephen Hawking's: "Black holes ain't as black as they are painted. They are not eternal prisons as we thought. Things can get out of a black hole, both to the outside and possibly to another universe. And so, if you feel you are in a black hole, don't give up. There is always a way out."

This was an encouraging end to a momentous press conference. But, to stick with Hawking's thought, do black holes really give us a chance at resurrection after we've gone through hell? Are black holes just a temporary state of purgatory on the way to true epiphany?

Let us not be led astray by this deceptive hope. The probability that after I die and am cremated a cyclone comes along and sweeps together the leftover ashes and smoke and I am reconstituted from all these scattered remnants is far greater than my ever coming back after being swallowed by a black hole.

Theoretical physicists aren't satisfied with this practical impossibility, however. The bare prospect of such an occurrence causes them no end of consternation.

INFORMATION LOSS

Every age has its great topics. They influence our conception of the world, and they influence science as well. A colleague of mine once made the sardonic observation that it didn't surprise him that the term for the beginning of the universe, the *Big Bang*, was chosen shortly after the detonation of the first atom bomb. Today we live in the Information Age, and more and more often now we see physics being rewritten in the language of information theory. The most modern variations go so far as to claim that gravitation can be described in bits, that natural laws resemble programming language, or even that the whole universe is in reality just one big computer simulation.[4] I don't really find such wild speculation convincing, but it certainly is the case that information has become an important concept in the natural sciences.

Everything is information: matter, energy, maybe even black holes. One of the most crucial concepts here is the concept of the opposite of information, noninformation, disorder, or to put it in lofty terms, entropy. The fact is, the concepts of light and time, knowledge and ignorance, chance and fate—they're all closely related.

Back in the late nineteenth century the Austrian scientist Ludwig Boltzmann was investigating the relationship between aspects of thermodynamics—for example, heat, pressure, energy, and work—and the smallest particles. In Boltzmann's time, the heat and pressure of steam engines produced energy and work. In a steam engine it is many small water-vapor particles that move and create the pressure that does the work and powers a locomotive.

The particles in the boiler are like children playing in a bouncy castle. The more wildly they jump around, the more violently the bouncy castle is shaken. The pressure on the walls increases the more children are bouncing around within them and the more un-inhibitedly they jump. The energy and velocity of the individual children correspond to the temperature in a boiler. By the end of the birthday party, the children have tired themselves out and the energy starts to ebb. The bouncy castle comes to a rest; the boiler cools down.

Before the bouncing begins I divide the children into two groups: to start with, the calm children in blue T-shirts are sitting down inside the bouncy castle. When the start whistle blows, the wild, athletic children in red T-shirts rush inside and there are a few major, but for the most part bloodless, collisions. When the wild horde makes it inside, the castle starts to wobble alarmingly as almost all of them make a coordinated and simultaneous lunge at the back wall. At this point the prevailing condition is still one of high order and low entropy. Because the castle is so full of children, however, the calm children have to start jumping, too—otherwise they'll get knocked over—and the wild children have to jump a little less wildly—otherwise they'll keep knocking into each other. Both groups become mixed; the mayhem becomes more generalized and less easy to differentiate. A physicist would describe it in the following terms: the bouncy castle has reached thermal equilibrium, and the entropy has increased— soon everything is mixed together, red and blue T-shirts everywhere. If the kids were to take their shirts off, no one would know which group they'd started in.

Something similar happens in a steam engine. If I connect a boiler full of hot air to a boiler full of cold air, then the air from the hot boiler will flow into the cold boiler and I can power a turbine. If I stop adding heat, the temperature in both boilers will even out, the gas particles in each will start moving at equal speeds, the air will no longer flow in just one direction, and the turbine will come to a stop. The system has reached thermal equilibrium; all the particles are completely mixed. An orderly system—cold particles over here, hot particles over there—has become a disorderly system; the entropy has increased; it no longer does any work. The physicist says the system is thermalized—meaning it is now completely mixed. There is only a large, assimilated mass, whose single characteristic is the common temperature of all particles.

We could also say that the disorder only ever increases. This is one of the most important discoveries young parents make, and the same concept applies in physics. It describes a basic principle of thermodynamics, which holds true in every closed physical system and in every child's playroom: no one will ever witness two boilers with the same temperature spontaneously dividing themselves into one hot and one cold, just as the toy blocks in a child's playroom will never organize themselves by color. You must always apply energy first in order to decrease the entropy. Tidying up is tedious—and costs energy.

Nevertheless, not even a disorganized box of different-colored toy blocks has reached the maximum level of entropy possible. This is reached only when all the toy blocks are ground up, disintegrated, and finally have radiated off into diffuse thermal light. So, even in a child's untidied playroom, things can always get worse.

We can consider ourselves lucky that our universe is only a few billion years old. If we lived in an infinitely old universe, then despite all our efforts it would be characterized by maximum disorder and complete randomness. There would be no more galaxies, no more stars, no more particles, and no more black holes. Light would be infinitely stretched out and seemingly extinguished. The universe would be about as exciting as the smoke of a blown-out candle in the desert wind. In this sense the finite nature of the universe is ultimately the prerequisite for our existence.

Interestingly enough, the concept of entropy also appears in information theory, as the American mathematician Claude Elwood Shannon demonstrated back in 1948—you just need to replace the toys in the child's bedroom or the gas particles in the boiler with letters. Let's take the pages of this book. If I play telephone and quietly read these lines to my neighbor, who quietly repeats them from memory to her neighbor, who in turn repeats them to her neighbor, then more and more errors will creep in the longer the chain is. What is hopefully a halfway informative text will eventually turn into incomprehensible gibberish that no longer contains any information at all. If I just continue to relay information without making corrections, information loss and disorder will continually increase. A hot bowl of alphabet soup will never, within any conceivable span of time, become a comprehensible book.[5] What's needed is energy, purposefully applied—for example, in the form of solar energy stored in chocolate, which the author's brain needs in order to write logical sentences.

The concept of entropy can be extended to black holes. They are in fact the ultimate equalizers and destroyers of information. Einstein's laws, applied to a person who falls into a black hole, state that all the information within him, his whole history, his thoughts, his appearance, his gender, and his memories will be reduced to a single number: namely, how much he weighed at the moment of his departure from this universe. Thus, five sandbags would leave more of an impression on a black hole than the president of the United States.

The whole system formed by a black hole is entirely defined by its mass and its angular momentum. In this respect black holes are, despite their monstrous size, the simplest and plainest objects in the universe. Every single cell in an earthworm is immeasurably more complex than a black hole.

If black holes really have a Hawking temperature, then it can be demonstrated that the surface of their event horizon is a measure for their entropy. Because black holes, according to Einstein's theory, can only get bigger, their entropy can only increase as well, while the total information, the total complexity of the universe must decrease. Any time a person or even an earthworm is lost, the universe also loses a tiny bit of history. On Earth at least, they would leave behind their mortal remains, but if the person or worm disappears into a black hole, the loss would be total.

If Hawking is right, then black holes eventually evaporate, meaning that their mass, size, and entropy decrease. The total entropy in the universe wouldn't decrease, however, because the emitted radiation would carry the entropy along with it. For a person swallowed up by the hellish maw of a black hole and reduced to a single point, that means that ultimately he will be split up into the tiniest of individual components and radiated out to all parts of the universe.

All his thoughts would make it out somehow, true, but they would be hopelessly jumbled up and inaudibly mixed in with the quantum static of the universe. Given the inexorable expansion of the universe, they would finally be lost to nothingness.

An evaporated black hole would thus be like a tipped-over box of different-colored toy blocks—a complete mess. But because the total entropy shouldn't change as a result of the evaporation, that means that black holes are already a complete mess to begin with. And indeed, today almost the entire entropy of the universe is contained in black holes.[6]

Many theoretical physicists can't come to terms with this information loss, however, and they speak of the information paradox of black holes. In quantum physics, the preservation of information is sacred. Only when all the information is preserved can we certify that a quantum system develops lawfully and predictably. The condition of an unbothered, unmeasured, unseen quantum particle at a given point in time is determined precisely by its prior condition.[7] The present and future of a particle are thus clearly linked. The equations of quantum physics are reversible: you can run them forward and backward and always get the same result. In quantum physics, however, the condition of a particle is always just a measure of probability that can be assigned to *one* value of a particle with relative accuracy—but the other value remains undetermined. According to the Heisenberg uncertainty principle, the values of a particle can never be measured exactly, and every measurement of a particle can in turn change the state of the particle.

You can imagine it like archery: when a good archer aims at the target, you can be relatively certain that she'll hit it. But there's no way to predict exactly which ring the arrow will end up stuck in—this can only be said within a certain probability. Only when the arrow hits do you know the exact number of points awarded for the shot.

Quantum particles are like arrows flying through the air. Once they're measured, the arrow has hit the target. If you look back, you can also say which archer shot the arrow. The problem is reversible— arrow and archer are linked. Because of this, physics is capable— within a certain margin of error—of making impressively exact predictions, and of linking together causes and effects.

But if black holes destroy quantum information, then they also interrupt the clear path through time. The flight of the arrow would be interrupted, so to speak. We wouldn't know where it came from or where it was going. It could suddenly hit anywhere—maybe even next to the observers who were standing behind the archer. A crack in the dogma of informational preservation casts doubt on the omnipotence of quantum physics and the predictive power of physics in general—no small problem.

A few theorists think that maybe all quantum information is stored in the center of the black hole, that is, in the vicinity of the singularity. But then all the information that has ever disappeared behind the event horizon has to remain there until finally the black hole evaporates. This doesn't make much sense, though, because even to store information requires space and energy. Ultimately the black hole is so small that it simply doesn't have room to store the information of a billion suns.

Other physicists suggest that the information lingers on or just beyond the event horizon. Could it be that when something crosses the event horizon, the latter vibrates like a membrane and stores information in that way? Is it possible that black holes are merely information, stored on their surface? Einstein would turn in his grave if he heard either guess, because according to his equivalence principle a particle in free fall that drops into the dark abyss of a black hole shouldn't even notice that it passes the event horizon. Only when it smacks into the singularity should it realize that something's gone wrong. In the theory of relativity there's no room for information on the event horizon.

Nevertheless, most physicists assume that black holes store information in one way or another and release it again as they give off radiation—and that the radiation black holes emit even contains a secret code, which, at least in theory, it might be possible to decipher in order to learn things about their past. Stephen Hawking himself, after initial doubt and a lost bet,[8] eventually went over to this school of thought. Meanwhile the famous mathematician and black hole pioneer Roger Penrose insisted that information was truly and irrevocably lost inside black holes. We simply don't yet know what gravitational fields actually do to quantum particles.

I tend to lean toward Penrose's position, though with some caution. Black holes are macroscopic objects; they aren't just limited to the singularity at their center—rather, a black hole is the entire area of curved space-time around it. It is collectively made up of all the quantum particles in and outside of the singularity. No quantum particle is isolated and unaffected by the others. Information is collectivized.[9] Given that this is the case, can you still talk about individual relationships and the information in individual particles? If space isn't quantized, does it make sense to argue about space using the principles of quantum physics? Quantum theory is reversible, but the real macroscopic universe is not. Why should black holes be? Maybe they're the biggest random generators in the cosmos?

Physics is in the midst of an information crisis; entire books are written on the subject. Which is wrong: the theory of relativity or quantum physics? There are a lot of strong opinions, but we don't know which of them will lead us anywhere. A crisis in physics is always a chance for a new theory, however. Physicists have been searching for it for more than forty years—but so far without success: we still can't get gravitational and quantum physics to work in harmony together. Developing a theory of quantum gravity is an unfathomably complex undertaking. Most of the attempts have the utmost difficulty just getting an apple to fall to the ground.

It's not creative ideas that are lacking, though, rather a clear God-given hint as to which of these ideas is the right one. One of the leading minds in this field, the quantum gravity researcher Hermann Nicolai of Potsdam, once said to me: "I don't think we're going to get any further with thinking alone—we need the experiment." We need an Eddington expedition for quantum gravity!

So far though, this crisis in physics has been primarily of a theoretical nature. Our image of the black hole doesn't yet put us in a position where we're able to confirm or rule out many new theories. At the moment, the theory of relativity is all we need to understand it and some parameters of general relativity are better constrained than with any other method. If a new theory makes a difference of a few percentage points in the size and shape of the shadow, maybe we will eventually be able to actually see its effects. If the deviations only occur at the scale of quantum objects, they might just remain hidden to our eyes forever.

Thanks to the image of the black hole, the problem of the two theories' irreconcilability has now become a bit more real and tangible than it was before. If we look into the darkness of the shadow, then we are looking right at the edge of the event horizon, where relativity and quantum physics vie for supremacy. The problem of unifying the two big theories is by no means abstract. It's very real. What we've done is to give this problem a place, so that you can now point your finger at it. The true mystery of this image doesn't reside in its bright, fiery ring, but rather in its shadow.

OMNISCIENCE AND LIMITATIONS

IS EVERYTHING MEASURABLE?

One of the Hubble Space Telescope's most captivating images was taken around Christmas 1995. For ten days the telescope stared at an unremarkable and almost randomly selected patch of sky just above the right edge of the Big Dipper and took 342 individual photographs. Combined into a single image, they produced the Hubble Deep Field. Compared to the vastness of space, the section it depicts is very small. It is about as large as the view of the sky you'd get through the eye of a needle held at arm's length. The image is teeming with larger and smaller bits of light that are scattered around the darkness of space. If you look closer, you discover that each one of these little specks of light is its own galaxy—3,000 of them in a single image. In order to depict the entire sky, you'd need about 26 million of these needle's-eye-size images, which would give you a few hundred billion galaxies. Considering that each galaxy is home to hundreds of billions of stars, that means our universe contains at least 10^{22} stars—though probably a lot more than that.

"As the host of heaven cannot be numbered, neither the sand of the sea measured"[1]—so wrote the prophet Jeremiah over 2,500 years ago, expressing the idea of immeasurable size. Although he could only see a few thousand stars with the naked eye, he already had a sense of the incomprehensible vastness of space. There are in fact about as many stars in the sky as grains of sand on the beaches of this world—though the latter's number is much more difficult to precisely determine.

We are living in a special time. Today we can see with our own eyes what the prophets of old had only the faintest notion of. Telescopes and satellites open up a glimpse into unknown worlds, an opportunity granted to no other generation before us. Like God himself, we can look down on the Earth from above and see it floating, a blue pearl set against the black velvet backdrop of space. We see clouds and sandstorms on Mars and giant glowing dust clouds from which new stars are born. We see distant galaxies with hundreds of billions of stars in every color of the rainbow, and a needle's-eye section of the sky filled with thousands of galaxies that conveys only a small fraction of the abundance that populates the universe. The wealth of images from space surpasses anything a single person could take in and understand, and our knowledge keeps growing and growing.

This represents the clear success of science and technology. This is our age—the era of natural science. Everything is measurable— even human beings. Where before, intuition, hope, and faith helped us to make decisions, now we have studies, measurements, models, and data banks. Every decision is supposed to be rationally justified and backed up by data and models. Today even theologians and scholars in the humanities work with computer-based methods and statistical approaches that are borrowed from the natural sciences. Technology has our lives fully in its grip, and on top of that offers entertainment and inspiration. We no longer play in the yard, but on the computer. God has been tamed; humans have become predictable. Will we eventually reach the point where we're able to find a rational and scientific basis for every decision we make—maybe with the help of an app?

Physics is far up at the forefront of this development. Physics and astrophysics don't just present us with the beauty of the universe; they also lead us to the very big questions in life. With our telescopes we look back to the beginning of space and time and investigate the Big Bang. And now we also look into the void of black holes. Who would have thought this possible just a short time ago? The beginning and end of time have come into view—is this not physics' greatest triumph? Is this not the latest step in a long evolution toward the complete illumination and comprehension of the world? Scientists on every continent are working together to solve the last big mysteries facing humanity. Who or what can stop us now? What secret can escape our gaze now that the whole world is working together?

It took a continent to discover the Higgs boson, to which we all owe a bit of our mass. It took research stations and researchers on two continents to discover the trembling of space-time and detect gravitational waves. And it took the entire world to finally make a black hole visible.

The world is preparing itself for the final battle to unravel the big questions of physics and of life. Is it just a matter of time before we've also solved nature's last mysteries? Before we can tear the veil of obscurity from the very face of God?

Throughout the history of science our horizons have continued to broaden, and knowledge and discovery have grown exponentially. What was a country became a continent; a continent became the entire world. The globe became a solar system, a solar system an entire galaxy, a galaxy an entire universe. Now physicists have even begun to speak of several different universes, the multiverse.

The German physicist Philipp von Jolly entered the annals of history with his remark that almost everything in physics had been discovered already. In the late nineteenth century, he advised the young Max Planck to turn away from the study of physics. "In this or that corner there might be some stray bit of dust or some small bubble to check over and classify, but the system as a whole stands rather secure"—so ran his advice, as Planck remembered it. Planck, then a student entering university, wouldn't be discouraged. He opened the door for Einstein's theory of relativity and launched quantum physics.

And so things will just continue in this same pattern, won't they? Is that really the case, I ask myself—and I'm not the only one.[2] Maybe the next big discovery is that we can't discover everything. To discover our limitations is also to discover humility.

Indeed, the new physics is actually based on the limits of knowledge, which have become a fundamental part of physics itself. The fact that in the theory of relativity the speed of light is finite means that we cannot know everything, cannot count every star in the universe, cannot measure everything exactly, and cannot predict anything perfectly. Quantum theory, with its Heisenberg uncertainty principle, leads to the realization that nothing exists in exact terms. Thermodynamics and chaos theory lead to the insight that the future is ultimately, and indeed intrinsically, unpredictable.

As of today, using the most modern methods, we have counted close to two billion stars out of the total in our Galaxy. This is nothing compared to the actual number of stars in all the galaxies in space. We will never manage to catalog all the stars, much less to visit them. They are only an echo from the past. Many don't exist at all anymore—we only see the light they sent us many ages ago. As a result of the continuing expansion of the universe, we could never even reach 94 percent of the galaxies we see today, even if we could travel at the speed of light.[3]

According to everything we know and according to present-day consensus, the Big Bang and black holes are scientific reality—but along with this reality, the limitations that they present us with have also become real. Everything beyond them continues to be the realm of imagination and mathematics. We cannot look into a black hole and we cannot listen back to the time before the Big Bang.

Naturally we will continue to push at these limits and search for windows into the previously unexplorable, but there is no guarantee that these windows exist. Any fundamental broadening of our horizons would require a radical revolution in everything we've come to understand about physics. Will physics let something like that happen? Whatever big momentous words we might use to describe its history, when looked at in hindsight, science appears to develop in one long process of evolution, rather than as a result of several revolutions: Einstein didn't make Newton superfluous; in a certain respect he only pointed out the limitations of his theory and embedded it within a new, more comprehensive one.

Revealing the last big mysteries of physics will require a combined effort from the whole world. Exciting things await us for many decades to come. But—what if it takes even more than that? Gigantic interferometers with dozens of colossal telescopes scattered across space? Particle accelerators of planetary dimensions? Who's going to pay for all that? Is it even doable? And even if it were doable, would it answer all our questions?

Maybe, I ask myself, the greatest triumph of the natural sciences is also their greatest defeat? Maybe in the midst of the final battle for total conquest and complete understanding of the world, we will realize that in our hubris we have chased after a mirage, and that on its own science won't bring us even one step closer to the answer to life's big questions.

Could it be that the big questions—where did we come from? where are we going?—can never be answered through the use of technology, and that we are prey to the delusion that everything is feasible? This wouldn't mean that we should stop asking questions— but we should approach nature, God, and questions about our existence with more humility.

There's still a lot of joy to be had from the great efforts of science to come, but this isn't a sacred goal in and of itself. Science is not an absolute method for explaining everything in and beyond the world, but rather a celebration of human creativity and curiosity. Ultimately we physicists will probably lose the final battle to answer the big questions—but nevertheless, the fight to bring light into the darkness is worth it.

THE FOG OF TIME

Science has seemingly prophetic powers—it makes amazing predictions! This power of prophecy is a key demand that science makes of itself, and to many people an impressive achievement. The flight of a ball, how different materials behave, the way light is deflected near the sun, or the appearance of black holes—all can be calculated wonderfully in advance. Today even weather forecasts have become quite useful, and scientists are working at full steam on prognoses for the course the coronavirus pandemic will take. Will we someday reach the point where everything is predictable, and at that point will everything in the future already be predetermined? Our intuition balks at such a thought—and, thankfully, is right to do so.

When I was young I used to wonder what time was. I imagined

time as a forest, wreathed in thick fog, which I had to walk through without stopping. Only God can look down from above and see all the possible paths in this forest of fog; only he sees past and future at the same time. I myself, however, see only a small stretch of path before me and a small stretch of path behind me. Before me the future emerges bit by bit from the fog of uncertainty; behind me the past fades in the fog of my memory. Sometimes I hurry through the forest, sometimes I move slowly, but I can never stand still. At every fork in the road I face a new decision. In this way my path changes and leads into a new, uncertain future. Other people travel their own paths in the forest of fog; sometimes I encounter them, sometimes we walk together, sometimes we lose sight of each other.

But why does our journey through the forest of fog only proceed in one direction? Why does time in real life only ever go forward? Why does the arrow of time only ever point in one direction? And why is our gaze into the future limited?

In space, after all, we can move forward and backward, left and right, even up and down. And in space we can return to the same spot again and again. Time doesn't work that way. Time is a parameter in many of the equations in physics, and in these equations you can fast-forward or rewind time like in a movie. In real life this isn't possible, however much we'd all like to turn back the clock sometimes.

In order to understand these questions, you have to look at every area of physics at once: the theory that deals with very small things, quantum physics; the theory that deals with very large things, the theory of relativity; and lastly, the theory that deals with many particles, thermodynamics.

One thing is clear: without time, there's no development. Time is both a blessing and a curse. To it we owe our birth and our experiences, as well as our decline and our death. To have time is to have a beginning and an end. In a static universe there would be nothing to suffer or to lose, but also nothing to experience or discover.

In general the emergence of time in physics is interpreted as a consequence of entropy—of inexorable decline. Unlike many other laws of physics, the second law of thermodynamics, the law concerning entropy, only moves in one direction: entropy must increase. Just like time. As a result, processes become irreversible—they travel in only one direction in time. Once you've burned a book to power a steam engine, the same book is never again going to emerge spontaneously from the ashes. Anywhere work is performed, where energy is applied, a small bit of energy is lost and disappears in the form of growing disorder. As we live, breathe, and move, we use up energy and increase entropy. Thus, anyone living can only move in one direction in time.

Gravity, too, is a strange one-way street. Electric charges can be positive or negative, can attract or repel; magnetic fields have a north and a south pole; only gravity has no counterpart. Mass only ever attracts. An apple in the Earth's gravitational field only ever falls downward, and, a black hole only ever gets bigger.

But it is precisely this one-way traffic that makes development possible. If there hadn't been any gravity after the Big Bang, then gas and other matter would have been lost to the emptiness of space. Stars and planets would never have coalesced; humans would never have evolved. Without gravity the sun wouldn't burn, plants would not grow, and people would not eat. We owe our existence to gravity.

The depressing law that states that entropy only ever increases also has a positive corollary: by means of the targeted application of energy, I can diminish the entropy in specific places. With a bit of energy I can tidy up the child's playroom, with a bit of energy I can write a book—at the expense of the total energy in the universe. Only the arrow of time and gravity allows for the existence of islands of creativity in space. The big question is only: Where did all this energy originally come from? This remains one of the biggest mysteries of our universe.

But the very thing that gives us life sets a limit to our desire to know everything. The greater the entropy, the less I know about the past or future of individual particles. I know that ultimately a burning book dissolves into ashes. But it's impossible to predict how the ashes will be distributed. The course of the world is thus inherently indeterminable and unfixed.

Sometimes in conversation I get the impression that there are still many people, their thinking shaped by science, who in their hearts— and against their better judgment—maintain a strictly deterministic worldview. If you could only know the exact condition of the world at a precise point in time, then the course of all things would be perfectly fixed and predictable—yes, it could even be calculated in advance. The world would indeed be one big computer game. The free will of each individual person would also just be an illusion—the result of the ultimately predetermined development of the quantum system of our brain cells under the influence of information that we receive from our environment. But then every decision you make would have been determined in advance—in fact long before you were even born. Did the Big Bang decide that I would now lift my finger in warning?

The world is not predictable; indeed it is fundamentally unpredictable! Physicists are very proud of all the things they can calculate, and rightly so, but sometimes they overlook their own limitations. For physicists, determinism is like a pink unicorn: in their dreams it is fascinating, but in reality it doesn't exist. Determinism can only be approximated over short spans of time and within small, limited spaces. If I line up dominoes with the right distance between them and knock the first one over, it's deterministically fixed that the last domino will fall over, right? But fundamentally, neither future nor past can be predicted. The fog of chance denies us a clear view into eternity. In real life the dominoes don't always fall how we think they will—for example, when Schrödinger's cat happens to come slinking into the room.

My colleague Simon Portegies Zwart of Leiden has demonstrated this in impressive fashion. He used a computer to simulate the movement of three nonrotating black holes, point masses, using only Newton's classical laws of gravitation, though numerically calculated to a random specificity. This is pretty much the simplest physical system you can imagine. You would expect that the development of these three gravitational systems, both forward and in reverse, could be predicted as long and as accurately as you like. In fact this is not the case, because within a span of time equal to the age of the universe, the system can change unpredictably if you don't know the distances between the black holes to within a Planck length of accuracy. A Planck length equals about 10^{-35} meters; it is the smallest distance that we can even know—far smaller than any quantum particle. It is fundamentally impossible to measure a distance that size, because at such minute dimensions all the known laws of nature fail to apply. This means that even in a system made up of three point masses, the development becomes irreversible and unpredictable. Conversely, such a system can't be traced back to its origins, either. What the beginnings of these three black holes were like, we cannot know.

If Simon Portegies Zwart and his colleagues had calculated a system of mutable planets instead of black holes, or instead of Newton's simple gravitational laws had used the more complicated equations of Einstein's theory of relativity, the system would have developed even more chaotically. If you add more stars and black holes, it all starts to look like utter chaos. We must learn to come to terms with the fact that the universe is fundamentally unpredictable and chaotic!

Do I have to add that a person is unfathomably more complex than a system of three black holes? Probably not! Even within a short time frame, no person is predictable, as all parents of small children know. Thus, anyone who dreams of transferring humans onto a computer at some point and calculating every aspect of their lives would do better to dream of pink unicorns—unicorns at least aren't physically impossible. It's true that humans are subject to natural laws, but they themselves, at the most profound level, are quite fundamentally free!

Even at the microscopic level, the origin of a decision in our brain is very quickly lost in the fog of uncertainty. But it isn't the foggy quantum froth in my brain that makes my decisions for me. Contrary to the claims of some physicists, I do still have free will and am therefore not free from responsibility for my actions.[4] I cannot fob this responsibility off on the quantum particles in my brain, claiming that they have nothing to do with me and make arbitrary decisions on my behalf, because "we" aren't as chaotic as that. I'm not just the sum of the individual components into which I can be divided; I am also the interaction between them and their development over time. From all this, something new, something autonomous is constantly growing—me, myself.[5]

What this self is, however, is as vague in philosophy as the nature of time is in physics. Part of my conviction is that I am comprised not only of the quantum froth in my brain, but also my past and my future—as far as my horizon reaches. Gathered within myself are my thoughts, my memories, my present, my hopes, and my faith. I am all of this. Thus, I can change as I go through time, because my horizon shifts with every step I take; it moves with me. I am therefore constantly changing, without ever completely becoming someone else.

But where, speaking in terms of physics, does this fog of time come from, this uncertainty in both directions? The reason why we can't accurately look ahead or trace things back rests on the fact that we cannot know anything in this world with absolute precision.

By way of example, in order for us to know something with infinite precision, then we would have to measure it infinitely long—but this can't happen in a universe that has a finite age and a finite size. Fundamentally speaking, in a universe with time, nothing is exact. If something is infinitely small or infinitely short, you would have to apply an infinite amount of energy to measure it with infinite precision. This can even be mathematically proven,[6] and leads to the famous Heisenberg uncertainty principle, which states that you can never precisely know all the values of a quantum particle—and what you fundamentally cannot know with precision doesn't exist in precise physical terms!

In this sense, the mathematical equations that we learned in school are deceiving. They describe a nature that doesn't exist—not in that exact a form. The Swiss physicist Nicolas Gisin has therefore suggested we use a new, intuitive mathematics[7] that takes into account the inexactness of numbers. Only with time do numbers become more exact. To state it in exaggerated terms: "two plus two equals four" is only exactly true after an infinitely long period of time. In order to find out, for example, if a loaf of bread weighs exactly two kilograms, I would have to measure for an infinite amount of time, but by then it would have long since gone moldy—or been eaten.

If the speed of light were infinite, all the information in space would reach me instantaneously, even if we were separated by an infinitely great distance. The knowable universe would have no limits and would be infinitely large. Everything would be connected with everything else simultaneously. But because the speed of light is finite, there is no knowable infinity in space or time and therefore never an absolute precision. The finite nature of the speed of light thus grants us a particular freedom—only the here and now counts. Every place has its own unique present, past, and future. Right now I can't yet know what will influence me tomorrow. Indeed, I can't even see it; I can only wait for it. The future will only be truly visible tomorrow—probably.

The finite is also what makes our lives possible in the first place. A limitlessly expanded and infinitely old universe would, according to the second law of thermodynamics, be endlessly random and eternally boring. If, after an almost infinitely long time, all the stars had burned out, all matter had disintegrated, and every black hole had evaporated, the universe would be an empty, unstructured ocean of radiation, full of infinitely weak light waves.

It is therefore the beginning that makes our universe so livable and lovable—and as the saying goes, there's a little bit of magic in every beginning. But we shouldn't be too afraid of the ending, either. In the development of the universe there have been so many surprising twists and so much creativity that we're entitled to expect a bit more yet. Why should the creative power that created a beginning not endure?

Life in our universe is finely balanced between the arbitrary and the predictable. We are neither free from natural laws, nor are we their slaves. If you're looking at a single particle, then the future is completely random. If you're looking at several particles over a set period of time, then everything happens with a certain probability and regularity. If you observe an extremely large number of particles over especially large periods of time, then just about everything is possible for every single one of them. Human life plays out in the middle realm: partly predictable, with a chance of chaos and sunshine, but also with the freedom to make new decisions again and again. It seems to me that the forest of fog is a metaphor that illustrates the condition of human life quite well.

IN THE BEGINNING . . . AND BEYOND

As a child I often lay awake at night, thinking. "What's behind the sky?" I asked myself. "And if something's behind the sky, what's behind that? And what's behind what's behind what's behind the sky? Is God there, or is it infinite heavenly emptiness?"

To ask these questions, some physicists claim, is childish.[8] But to ask questions like a child doesn't automatically make you childish! I'm happy to remain childishly curious and to never stop asking questions—I couldn't do differently if I tried.

I became a scientist in order to be able to see further, but my scientific gaze will never reach all the way into infinity. Infinity is something I can neither actually conceive of nor effectively measure, which is why the infinite is not accessible to science. Infinity is a mathematical abstraction and a metaphysical speculation.

In today's established model of the universe, our glimpse of the infinite ends with the Big Bang. With it begins our time and our history; everything that will ever be is contained within it. The Big Bang is an excess of concentrated energy.[9] Everything we see today—every form of matter or energy[10] and even us ourselves—can ultimately be traced back to this primordial energy.

An almost infinitely small space suddenly expands and grows exponentially in just 10^{-35} seconds. It's a primordial lightning flash of pure energy and light, out of which a quantum molasses of elementary particles crystalizes. Protons and electrons are formed, the building blocks of matter. After 380,000 years protons and electrons pair up and form hydrogen, which floods the universe. Matter and light are suddenly divided from each other and each go their own separate way. Dark matter becomes concentrated under the influence of its own gravity: dark galaxies arise out of the remnants of the Big Bang and gather the hydrogen around them. From this are formed galaxies with blazing stars that create new elements and fling them back out into space in giant explosions.

From the ashes of the first stars, new stars, planets, moons, and comets are born. The stellar life cycle begins, and in the end our Earth is formed. Water falls upon the Earth and collects there, and together with the stardust it forms fungus, single-cell animals, and plants. This new life changes the world, an atmosphere starts to form, clouds open up, animals evolve. The last to appear are humans, who under the light of the sun, moon, and stars populate and conquer the Earth, build cities, comprehend the world, time, and space, and write books about them—all of this thanks to the cosmic tohubohu of the Big Bang.

The fact that our universe even functions at all is incredibly astounding. To bring forth a universe is a tightrope act of physics. If the gravity were much stronger, stars would collapse into black holes. If it were weaker, dark energy would cause everything to scatter. If the electromagnetic force were stronger, stars couldn't shine.[11] That the gears of this cosmic mechanism all interlock and make life possible for us remains the greatest marvel of all time. If anyone had been around at the time of the Big Bang and predicted that she herself would be formed from all that chaos, she would have been called crazy. Physics textbooks simply do not allow for matter that suddenly starts thinking for itself, forming its own opinions and personality, and exercising creativity—and nevertheless, here we are.

A favorite answer that people like to come up with to explain this mystery is to say that in fact there isn't just one universe, but rather many universes that sprout up and wither again like flowers in a field—every universe a bit different. It's only by chance that we happen to live in this particular universe that fosters life, because this is the only one we can see.

Should we think even bigger then? Could we possibly discover traces of old universes in our own, for example, large-scale structures that were created by the collision of two different universes? Personally, I'd go so far as to guess that hypermassive black holes are potentially the best fossils of old universes—they are, after all, the last thing that should be left over from a universe like ours. To date, nobody has found any evidence of this. As yet, there have been no indications that parallel universes really exist and that we can measure them.

It's also not clear that one can conclude from the pure fact of the existence of one highly improbable universe that there must be many of them to make our universe more probable. If my neighbor wins the lottery, that doesn't necessarily mean that he's played millions of times.[12] The most we can say is that we happen to live next to a real lucky duck. If this were the one drawing we had ever experienced and we weren't exactly familiar with the rules governing it, we wouldn't be able to deduce how many lotto players—or how many universes—there were.

Without concrete hope of proof, the question arises as to whether the idea of the multiverse is physics or metaphysics. We can't look back through the singularity at the beginning of our own universe, nor can we look past the edge. Even if you argue that multiverses are real physics and not just wishful thinking, the question still remains: Where does the multiverse come from? All we've done is to shift our ignorance into the no-man's-land of physics!

Stephen Hawking claimed that to ask what was before the Big Bang made as much sense as to ask what was north of the north pole. He proposed models of the world in which the time coordinates never begin at zero.[13] To me this just seems like clever sleight of hand, because the north pole is only a problem in a certain model of the world and within a certain system of coordinates. Sure, someone who is only thinking of a world that is limited to the surface of a globe will in fact be unable to answer this question. Nevertheless he can still proceed outward in every direction above the north pole and ask himself what might be above or below it.

Others say that the universe was formed spontaneously from nothingness, but that depends on how nothingness is defined. Every theory of how the world was formed begins with a set of natural laws, a set of mathematical equations—and today, in most cases, with an ocean of diffuse quantum foam from which a new universe can spontaneously emerge. In no model does a universe actually form from nothingness—and the same holds true for several universes.

"In the beginning was the Word..." So begins the opening verse of the Gospel of John, one of the most famous verses in the Bible.[14] At the beginning of every branch of the natural sciences are the rules according to which the world functions and from which a "language" is constructed. But where does this word that was at the beginning come from? Where do the rules come from? That which, with the help of the rules, becomes *something*—where does *it* come from?

"... and the Word was God," says the second, key part of the verse. People have been asking about the first cause, the prime mover, for millennia, and in the Judeo-Christian-Islamic cultural sphere the answer to this ancient question is "God." "God" is in a sense just a placeholder that everyone must fill in for themselves. The key question that then presents itself is: Who or what is God? Even the way the question is formulated makes it plain that here we are touching on a dimension that goes far beyond physics and its limitations.

Nevertheless, a person might decide that the question of God isn't a topic for physics to address. For an individual to take an agnostic stance is entirely understandable. How one deals with the question as to what is the origin, what is the meaning of life remains a deeply personal decision. One doesn't have to ask it, but one may.

An agnostic position can also be perfectly reasonable, given the background of how modern astrophysics developed. Astrology and astronomy split off and separated from each other over the course of a long process that stretched from antiquity into the modern age. Today an astronomer who practices astrology wouldn't be taken seriously as a scientist by his colleagues. He would be accused of sham science.

The process by which the sciences came to assert their independence has led in the modern age to the complete exclusion of religious, philosophical, and theological questions from the natural sciences. This was part of an emancipatory process, whereby science freed itself from the dictates of the church and of philosophy. But that doesn't mean that one should fundamentally disregard these questions. Restricting itself to nonreligious questions is a method chosen by science; it is a not a universal answer.

By the same token, one can't use science to deduce that there is no God—simply because one doesn't admit the question as to God's existence into physics. Atheism is a legitimate belief, but it has no scientific basis. The attempt to use science to disprove the existence of God strikes me as being just as absurd as the attempt to use science to prove God's existence.

It isn't just black holes that show us that limitations are part of our world. If you dare to ask questions that extend beyond the bounds of physics, you won't be able to get around the question of God. For the very reason that nature places fundamental limits on what we can know, we run up against these limits again and again, and with our questions we rattle at the gates of heaven. There is also a kind of solace to be found in limits, because they frustrate human arrogance and allow us to believe and to hope. I don't think that a completely godless physics is possible, not if you're truly asking questions that go right to the limits of human knowledge—and then continue beyond these limits. We humans carry the big questions with us, deep inside. To ask where from, where to, and why is something like a primordial instinct, a part of our human soul. These questions occupy us for all our lives and send us searching. Religion, philosophy, and science each play their own role in this search. Things tend to get difficult, though, when one discipline claims the exclusive right to interpret the entire world.

Science would do well then to accept its limitations and be a partner in a constructive dialogue, instead of elevating itself to the role of ultimate explainer of everything. Otherwise science itself could all too easily become burdened with miraculous expectations and promises of salvation that it can't deliver on. I consider it dangerous to rely on science and technology alone to fulfill our spiritual needs— dangerous for us, but also dangerous for the credibility of science.

But is God even worth talking about today? Hasn't scientific progress shrunk God down to the role of a mere stopgap; hasn't our knowledge forced him into an ever smaller, ever more remote niche? To claim, as Stephen Hawking did, that God is superfluous because modern physics has already answered all the questions is making things too easy. On the contrary, I say that today God is more necessary than ever. Ultimately science hasn't come a single step closer to answering the big philosophical question of where it is we come from, even if we have uncovered countless facets of the development of life and the universe. Just as you can't really approach the infinite, so can you not really approach the origin of creation. Today we know far more than we ever have before, but we also know more of what we cannot know. The gap of unknowing that God is meant to fill has become larger and much more fundamental than it ever was. It contains the origin of the entire universe, or possibly many universes, and of the whole subatomic quantum world. Where did this all come from, and where is it going? We have come to better understand the rules of the game in the universe, but where the game and where the rules come from, this we haven't answered. We might place ourselves atop our Babylonian tower of knowledge, looking down at our little world, and proclaim the all-encompassing triumph of science, but we would be proclaiming a triumph that can never come to pass. We might declare that God is dead, but we would not be the first whose actions brought a mild smile to his face as he gazed down at us from afar.[15]

Therefore the debate between faith and science seems to me like the race between the tortoise and the hare. The hare, like science, underestimates his opponent, but when he reaches the finish line he sees that the tortoise is already there waiting.

But isn't God just an abstraction and a human projection? Certainly, because every notion of God is always human and abstract. Our minds try to make the incomprehensible comprehensible, and in order to do so we employ abstract concepts. But that doesn't mean that what peers out from behind those concepts doesn't exist. A complex number is an abstract concept found in mathematical equations, but nevertheless it led to the prediction of the very real positron, which actually exists.

In point of fact, natural laws are also abstract human constructs, though the processes they describe are entirely real. Strictly speaking, natural laws exist only in our heads. An apple doesn't know anything about the Newtonian laws of gravity or Einstein's theory of relativity, and still every apple falls downward every time, no matter what height you drop it from. The laws of gravity are real because an apple falls; for the same reason, God as the prime mover is real, because the world came into being.

Natural laws are abstract descriptions of reality written in the language of mathematics. But natural laws don't comprehensively describe the full extent of reality. They describe simple systems with astounding precision, but the more complex nature becomes, the harder it is to express in simple mathematics. Every mathematical formula, every computer program is always just an approximation of reality. Only reality itself is a perfect description of reality; only the universe is a perfect description of the universe; only a person is the perfect description of him- or herself. But we do not have access to this perfect description. Thus, all we have available to us are many insufficient points of access to reality, to the universe, and to ourselves as humans.

In the same way, only God is a perfect description of God. Any talk of God can only be inarticulate. Anyone who says he or she knows who God is or who God isn't has clearly not understood him or her. It is a sign of profound wisdom that in the Bible there is a commandment stating that we must not make any concrete image of God. God cannot be captured in any image. *Deus semper maior*—God is always greater than whatever we imagine God to be. This is just as true for believers as it is for atheists. It sometimes disappoints me to see how God is turned into a caricature, either to enlist him in one's own cause or to make fun of him. God is neither a flying spaghetti monster nor an old, white, clean-shaven American man.

But is there even any point in thinking about God? What's the use of talking about God if God lies beyond the horizon of what we can know? Even if we can't study the moment of the cosmos's creation, we can most certainly study its effects. Physicists do in fact make calculations concerning the interior of black holes, even though the interior can't be measured.

Gottfried Wilhelm Leibniz introduced a very limited notion of God in the eighteenth century, namely the image of God as a master watchmaker. God is the prime mover; in the beginning he set the world in motion, and ever since its perfect mechanism, which he built so masterfully, has kept on running, permanent and unchanging. God did such brilliant work that he no longer needs to worry about the universe. His world is the best of all possible worlds. Leibniz's God is the God of the Enlightenment, and he continues—unnamed and unnoticed—to haunt certain minds today, well knowing that the world isn't perfect.

In fact, even this idea of a watchmaker-God is not inconsequential, because in scientific thought the law of cause and effect is central. If I believe God to be merely the impersonal sum of all natural laws and initial conditions prevailing at the beginning of the world, then it's still these natural laws and initial conditions that determine the form and direction of our universe and that I measure. They reflect the beginning. The influence and the nature of God the master watchmaker would thus still be present and measurable today. Astrophysics, then, is in a certain respect a search for the lost traces of this master watchmaker in the light of the present.

In the same way, theologians have been racking their brains for millennia trying to figure out who or what God is, and searching for signs of God in the present day. For me personally, God is more than a watchmaker. In my religion, the Bible bears witness to a rich wealth of names, encounters, and stories of and about God. Other religions have comparable divine stories. These descriptions of God have developed over many generations from the joyful and painful experiences, questions, longings, and hopes of people living in this world. They all describe lived reality, but they aren't written in the language of mathematics, rather in the language of experience, of poetry, of dreams, of vision, and of wisdom.

The language of mathematics gives me no insight into questions like whether I am loved or what I am worth—except maybe if I myself am a mathematician who sometimes seems to be able to live from the beauty of math alone. To think that I could and should simply brush all this human experience aside, merely because today I understand the physics of reality better than people did a hundred years ago, seems to me presumptuous, if not downright arrogant.

Thus, the search for God remains highly relevant and important. The way I think about the beginning also determines how I look at today and tomorrow. From the watchmaker God, I expect consistency and reliability, but no interest in me or anyone else as a person. If, however, I believe God to be not just a something, but also a person, that is, someone like the God found in monotheistic religions, then I expect God to be someone with whom I can interact, someone from whom I can still expect new things today and tomorrow. Such a God makes encounters possible. In Christian belief the personhood of God is expressed in the symbolic suffering and sacrifice of Jesus Christ, as well as in the community of believers and the majesty of creation.

The idea of describing God as a person might lead agnostic or atheistic physicists to have their doubts about me, but the idea is less outlandish than you might think. Protons clearly seem capable of having personalities, since they can form a human being. It obviously is possible for physics to take a Big Bang, a bit of matter, and a few natural laws to bring forth humans with consciousness, who are capable of abstract thought, emotions, and humor, and have a sense of destiny and responsibility. The possibility that life, individuality, and personhood would form must therefore have been present in— though not necessarily preordained by—the laws of the Big Bang. It's obvious at the very least that this possibility wasn't ruled out, since, well, here we are! To loosely adapt Descartes's fundamental insight, "I think therefore I am," one could also say: "I am, therefore it's possible." If matter thinks and feels, why then shouldn't a creator God, the prime mover, be able to have a personality with a mind, emotions, and reason? If physicists can think up a cosmos full of life, limitless possibilities, and multiverses, then it seems to me that the idea of a personal God wouldn't be that irrational at all—in any case it's far more rational than understanding the world to be a preprogrammed computer simulation, as some of my colleagues secretly do. Just because many people for thousands of years have believed in a personal God, that doesn't automatically make belief misguided and outdated.

The personhood of God does, however, lie beyond any means of detection available to physics. If the scientific exploration of the universe has shown us how small we are, then God tells us how valuable we are. One's sense of worth is not a physically measurable quantity. It must come from without and be felt within. If someone declares their love for you, this declaration cannot be comprehended with particle accelerators or telescopes—unless maybe I were to consider the entire miraculous universe, even with its painful sides, to be one big declaration of love to us humans. A declaration of love is an extremely personal thing: one person might be fulfilled by it, another left cold. Two people who receive the same letter often read something quite different into it. To ask about the personhood of God is a profoundly human experience, which every single person must go through on their own—physics can't do it for us. And yet, these experiences can be shared; our own experience can resemble others'. Thus, they are not completely random or arbitrary.

I'm always surprised when people ask me how I reconcile science and faith. The fact is that I'm no different than many scientists who laid the foundation upon which our present-day knowledge rests. Nicolaus Copernicus, Johannes Kepler, Max Planck, Arthur Eddington, and many more prominent figures in scientific history were deeply pious people. Still today, I can walk through the halls of the Netherlands academy of sciences and discuss the latest surprises in quantum physics with one person and deep theological questions with another.

For me, natural laws are part of creation, just as I myself am. If an apple falls downward, in harmony with the laws of nature, then for me it is excellent physics, but also the expression of a reliable creator who is the same yesterday, today, and for all eternity. For other people, it's just an apple falling.

God for me also isn't just something, but rather someone. I experience this side of God in my own life, the lives of the people who came before me, and the lives of those around me. I experience God alone in prayer, in collective worship with the congregation, in contemplation of Jesus, and in the majesty and beauty of the universe. When I look out into space, then I'm not just looking at nature, at life, and the vastness of the universe, but also at what lies beyond it. Physics reveals new wonders to me, but it doesn't take away my faith; rather, it expands and deepens it. If I look upon Jesus Christ the person, I discover the human side of creation and of the creator. In this way I find for myself a God who is beginning and end, a God to whom I don't have to prove anything and cannot prove anything and in whom I am already home.

But just as skepticism plays an important role in the advance of science, so is doubt an important element of my faith. The experimental ground for faith is life, and therefore I must always subject my life and my faith to criticism. Maybe so many people have their doubts about the church today because some churches don't doubt themselves enough! The nature of the world and the nature of God will always be more complex than anything our limited reason can comprehend. Science without self-criticism is quackery, religion without doubt blasphemy, politics without uncertainties fraud. We can't know everything.

The limits nature places on us and our lack of knowledge are also what is magical about us, because our limitations turn us into seekers. It is the very uncertainty of this world that allows us to continue to make new decisions and ask new questions. How unappealing would science be if there were nothing new to discover? What would life be without questions? A life in which everything was calculated in advance? What sort of God would it be in whom you no longer needed to believe because you already knew everything about him? There's a good side to not knowing everything and not being able to prove everything. This is also a form of freedom, maybe even the basis for it.

Naturally I can't forbid God from allowing herself to be proven in this world and taking from me the freedom of believing—though I would be deeply disappointed by her!

And of course, maybe a person's true calling in this world and far beyond it is to keep asking questions and to keep searching. This is what sets us apart from the great majority of the universe. The limits to knowledge are both a blessing and a challenge. It is in the nature of the horizon that one can never step beyond it but can always broaden it. We broaden our horizons by continuing onward: thinking, questioning, doubting, hoping, loving, believing.

At the beginning of this book I took you along on a journey into space, past the moon, past the planets of our solar system, and out into the Milky Way, to burned-out stars and black holes. The journey into the universe is a relay race in which generations of astronomers have kept passing the baton of knowledge and opened up new realms. For me, this journey is not a campaign of conquest; it's more like a pilgrimage, undertaken to expand our minds and our spirit. In the end this journey leads us back to ourselves and our unresolved questions. The time has come, then, for us to stop being over-proud conquerors of worlds and go back to being humble seekers.

Those who search always carry the hope within them that they will find something. Every seeker is always, of necessity, a bearer of hope. When my colleague and TV presenter Harald Lesch gave the keynote speech at the hundred-year anniversary celebration of the German Astronomical Society, he was asked afterward about the significance of humanity and faith. He referenced the apostle Paul, who wrote about what remains—what abides—of man: "But now abideth faith, hope, love, these three; and the greatest of these is love."[16]

We humans are just specks of dust on a slightly bigger speck of dust in the immeasurable vastness of space. We can't cause stars to explode, we don't set the wheels of galaxies spinning, and it is not we who span the vault of heaven above us. But we can marvel at the universe and ask questions about it. We can have faith, hope, and love in this world—and this makes us stardust of a very special kind.

If today the Earth were to vanish from the solar system, if the solar system were to vanish from our Galaxy, if our entire Milky Way were to vanish from space, it wouldn't make any difference to the universe, and yet nevertheless the universe would be missing something very valuable, namely our faith, our hope, and our love—and our questions, with which, again and again, we bring new light into the darkness.

Acknowledgments

The idea for this book emerged after the image's publication in April 2019 from a conversation I had with science reporter Jörg Römer of *Der Spiegel* magazine. He had interviewed me and later accompanied me to a few speaking engagements. At some point we were sitting in a Vietnamese restaurant in Hamburg and discussing black holes, God, and the universe. He, the critical journalist, and I, the devout scientist, but both of us united in our curiosity and our fascination with science.

This book that we've worked hard to write together is meant to be like our conversation. We wanted to connect my small personal story with the big story of humanity's discovery of the universe, and to tell it in such a way that it's accessible to everyone. For that reason we told this story from my perspective: things that I've experienced personally, things that I've learned, a few anecdotes from the life of a scientist—from curious kid to established professor—and in just a few instances a short Bible verse that I find moving.

This book also tells the story of a part of my life that wouldn't have been possible without the love, support, and forbearance of my family. My wonderful wife is not only the best school administrator and partner imaginable; she also proofread this book. Any mistakes that might still be found in the book were introduced after the fact.

My colleagues Professor Frank Verbunt (Utrecht/Nijmegen), Professor Gerhard Börner (Munich), and Dr. Markus Pössel (Heidelberg), who reviewed the manuscript, rendered us—Jörg Römer and me—an invaluable service with their critical comments. Sara Issaoun thankfully checked the English translation.

Our agent Annette Brüggemann played a key role in the book's coming about and helped to develop the idea. Stephan Grünewald from the Rheingold Institute gave us space for planning. For the German edition, publisher Tom Kraushaar and our editor Johannes Czaja, along with all the staff at Klett-Cotta, saw us through the project with much professionalism and devotion.

Lastly, I owe my thanks to all my colleagues for their work both past and present, even if not all of them can be named here. Many of the names had to be moved to the notes, and the list there is likewise selective and incomplete. All the coauthors of our articles on the black hole are listed after these acknowledgments, but I could mention many, many more.

Jörg thanks his wife and his two daughters, who often had to do without him during the coronavirus lockdown. He also expresses his gratitude to his employer, *Der Spiegel*, for making it possible for him to realize this project. Last but not least, his close friends and colleagues helped with aid and advice.

I will be donating a large portion of my proceeds from this book.

Frechen bei Köln, September 2020
Heino Falcke

List of EHT Authors

Kazunori Akiyama, Antxon Alberdi, Walter Alef, Keiichi Asada, Rebecca Azulay, Anne-Kathrin Baczko, David Ball, Mislav Baloković, John Barrett, Ilse van Bemmel, Dan Bintley, Lindy Blackburn, Wilfred Boland, Katherine L. Bouman, Geoffrey C. Bower, Michael Bremer, Christiaan D. Brinkerink, Roger Brissenden, Silke Britzen, Avery Broderick, Dominique Broguiere, Thomas Bronzwaer, Do-Young Byun, John E. Carlstrom, Andrew Chael, Chi-kwan Chan, Koushik Chatterjee, Shami Chatterjee, Ming-Tang Chen, Yongjun Chen (陈永军), Ilje Cho, Pierre Christian, John E. Conway, James M. Cordes, Geoffrey B. Crew, Yuzhu Cui, Jordy Davelaar, Roger Deane, Jessica Dempsey, Gregory Desvignes, Jason Dexter, Shep Doeleman, Ralph P. Eatough, Heino Falcke, Vincent L. Fish, Ed Fomalont, Raquel Fraga-Encinas, Bill Freeman, Per Friberg, Christian M. Fromm, Peter Galison, Charles F. Gammie, Roberto García, Olivier Gentaz, Boris Georgiev, Ciriaco Goddi, Roman Gold, José L. Gómez, Minfeng Gu (顾敏峰), Mark Gurwell, Michael H. Hecht, Ronald Hesper, Luis C. Ho (何子山), Paul Ho, Mareki Honma, Chih-Wei L. Huang, Lei Huang (黄磊), David Hughes, Shiro Ikeda, Makoto Inoue, David James, Buell T. Jannuzi, Michael Janßen, Britton Jeter, Wu Jiang (江悟), Michael D. Johnson, Svetlana Jorstad, Taehyun Jung, Mansour Karami, Ramesh

Karuppusamy, Tomohisa Kawashima, Mark Kettenis, Jae-Young Kim, Jongsoo Kim, Junhan Kim, Motoki Kino, Jun Yi Koay, Patrick M. Koch, Shoko Koyama, Carsten Kramer, Michael Kramer, Thomas P. Krichbaum, Cheng-Yu Kuo, Huib Jan van Langevelde, Tod R. Lauer, Yan-Rong Li (李彦荣), Zhiyuan Li (李志远), Michael Lindqvist, Kuo Liu, Elisabetta Liuzzo, Wen-Ping Lo, Andrei P. Lobanov, Laurent Loinard, Colin Lonsdale, Ru-Sen Lu (路如森), Nicholas R. MacDonald, Jirong Mao (毛基荣), Sera Markoff, Daniel P. Marrone, Alan P. Marscher, Iván Martí-Vidal, Satoki Matsushita, Lynn D. Matthews, Lia Medeiros, Karl M. Menten, Izumi Mizuno, Yosuke Mizuno, James M. Moran, Kotaro Moriyama, Monika Mościbrodzka, Cornelia Müller, Hiroshi Nagai, Masanori Nakamura, Ramesh Narayan, Gopal Narayanan, Iniyan Natarajan, Roberto Neri, Chunchong Ni, Aristeidis Noutsos, Hiroki Okino, Héctor Olivares, Tomoaki Oyama, Feryal Özel, Daniel Palumbo, Harriet Parsons, Nimesh Patel, Ue-Li Pen, Dominic W. Pesce, Vincent Piétu, Richard Plambeck, Aleksandar Popstefanija, Oliver Porth, Ben Prather, Jorge A. Preciado-López, Dimitrios Psaltis, Hung-Yi Pu, Ramprasad Rao, Mark G. Rawlings, Alexander W. Raymond, Luciano Rezzolla, Bart Ripperda, Freek Roelofs, Alan Rogers, Eduardo Ros, Mel Rose, Arash Roshanineshat, Daniel R. van Rossum, Helge Rottmann, Alan L. Roy, Chet Ruszczyk, Benjamin R. Ryan, Kazi L. J. Rygl, Salvador Sánchez, David Sánchez-Arguelles, Mahito Sasada, Tuomas Savolainen, F. Peter Schloerb, Karl-Friedrich Schuster, Lijing Shao, Zhiqiang Shen (沈志强), Des Small, Bong Won Sohn, Jason SooHoo, Fumie Tazaki, Paul Tiede, Michael Titus, Kenji Toma, Pablo Torne, Tyler Trent, Sascha Trippe, Shuichiro Tsuda, Jan Wagner, John Wardle, Jonathan Weintroub, Norbert Wex, Robert Wharton, Maciek Wielgus, George N. Wong, Qingwen Wu (吴庆文), André Young, Ken Young, Ziri Younsi, Feng Yuan (袁峰), Ye-Fei Yuan (袁业飞), J. Anton Zensus, Guangyao Zhao, Shan-Shan Zhao, Ziyan Zhu.

Juan-Carlos Algaba, Alexander Allardi, Rodrigo Amestica, Jadyn Anczarski, Uwe Bach, Frederick K. Baganoff, Christopher Beaudoin, Bradford A. Benson, Ryan Berthold, Ray Blundell, Sandra Bustamente, Roger Cappallo, Edgar Castillo-Domínguez, Richard Chamberlin, Chih-Cheng Chang, Shu-Hao Chang, Song-Chu Chang, Chung-Chen Chen, Ryan Chilson, Tim Chuter, Rodrigo Córdova Rosado, Iain M. Coulson, Thomas M. Crawford, Joseph Crowley, John David, Mark Derome, Matthew Dexter, Sven Dornbusch, Kevin A. Dudevoir (deceased), Sergio A. Dzib, Andreas Eckart, Chris Eckert, Neal R. Erickson, Aaron Faber, Joseph R. Farah, Vernon Fath, Thomas W. Folkers, David C. Forbes, Robert Freund, David M. Gale, Feng Gao, Gertie Geertsema, Arturo I. Gómez-Ruiz, David A. Graham, Christopher H. Greer, Ronald Grosslein, Frédéric Gueth, Daryl Haggard, Nils W. Halverson, Chih-Chiang Han, Kuo-Chang Han, Jinchi Hao, Yutaka Hasegawa, Jason W. Henning, Antonio Hernández-Gómez, Rubén Herrero-Illana, Stefan Heyminck, Akihiko Hirota, Jim Hoge, Yau-De Huang, C. M. Violette Impellizzeri, Homin Jiang, Atish Kamble, Ryan Keisler, Kimihiro Kimura, Derek Kubo, John Kuroda, Richard Lacasse, Robert A. Laing, Erik M. Leitch, Chao-Te Li, Lupin C.-C. Lin, Ching-Tang Liu, Kuan-Yu Liu, Li-Ming Lu, Ralph G. Marson, Pierre L. Martin-Cocher, Kyle D. Massingill, Callie Matulonis, Martin P. McColl, Stephen R. McWhirter, Hugo Messias, Zheng Meyer-Zhao, Daniel Michalik, Alfredo Montaña, William Montgomerie, Matias Mora-Klein, Dirk Muders, Andrew Nadolski, Santiago Navarro, Chi H. Nguyen, Hiroaki Nishioka, Timothy Norton, Michael A. Nowak, George Nystrom, Hideo Ogawa, Peter Oshiro, Scott N. Paine, Harriet Parsons, Juan Peñalver, Neil M. Phillips, Michael Poirier, Nicolas Pradel, Rurik A. Primiani, Philippe A. Raffin, Alexandra S. Rahlin, George Reiland, Christopher Risacher, Ignacio Ruiz, Alejandro F. Sáez-Madaín, Remi Sassella, Pim Schellart, Paul Shaw, Kevin M. Silva,

Hotaka Shiokawa, David R. Smith, William Snow, Kamal Souccar, Don Sousa, Ranjani Srinivasan, William Stahm, Anthony A. Stark, Kyle Story, Sjoerd T. Timmer, Laura Vertatschitsch, Craig Walther, Ta-Shun Wei, Nathan Whitehorn, Alan R. Whitney, David P. Woody, Jan G. A. Wouterloot, Melvyn Wright, Paul Yamaguchi, Chen-Yu Yu, Milagros Zeballos, Lucy Ziurys.

Glossary

AAS (American Astronomical Society): Professional organization that publishes two important journals of astronomy.

Accretion disk: Rotating gas disk surrounding an object of great mass, which, like a whirlpool, transports magnetic fields and matter (plasma, gas, or dust) to the center.

AGN (active galactic nucleus): The central region of a galaxy that emits large amounts of radiation. The phenomenon is explained by supermassive black holes.

ALMA (Atacama Large Millimeter Array): Largest telescope working in the millimeter and submillimeter range. A network of 66 radio antennas in the Atacama Desert in Chile, approximately 5,000 meters above sea level.

APEX (Atacama Pathfinder Experiment): Twelve-meter radio telescope in Chile, near the ALMA Telescope.

Arc second: Angular unit. A circle can be divided into 1,296,000 arc seconds. A circle has 360 degrees, every degree has 60 arc minutes, every arc minute 60 arc seconds. Used in astronomy to express transverse distances or dimensions in the sky.

Astronomical unit (AU): Mean distance of the Earth to the sun—standard measure used in astronomy. 1 AU = 149,597,870,700 km.

Atom: Building block of matter that makes up our elements. Atoms are made up of heavy, positively charged protons and neutral neutrons, found in the nucleus, and one or more shells of light, negatively charged electrons.

Big Bang: Starting point of our universe, when matter and energy came bursting out of a tiny point. According to the current model in use among cosmologists, this happened around 13.8 billion years ago. The universe has been expanding ever since.

Binary star: System of two stars that orbit each other. In the Milky Way, every second star is located inside a binary or multiple star system. If one star collapses and becomes a black hole, it can slowly swallow the other and produce x-ray radiation (known as an x-ray binary).

Black-body radiation (Planck radiation): Universal radiation that every opaque object emits and that is dependent only on the temperature and size of the object. Stars and the cosmic microwave background emit this type of radiation.

Black hole: Object in space whose mass is concentrated to a tiny point. In the area surrounding it, the gravity is so strong that not even light can escape. Black holes are formed from the collapse of very massive stars after a supernova; they are also formed in the centers of galaxies, where they can be billions of times heavier than the sun and are considered "supermassive."

Cepheid variables: Pulsating stars with periods between one and one hundred days. The brighter they are, the more slowly they pulsate. By measuring their period of pulsation, it's possible to calculate their true level of brightness, or luminosity, and, by comparing this to the brightness as measured, to calculate their distance. The farther away a star is, the weaker its light appears to us.

CNSA (China National Space Administration): Chinese government space agency, responsible only for satellites and space probes, not, however, for manned space travel.

Cosmic microwave background (CMB; 3K radiation): Blackbody radiation from the early phase of the universe, when it turned transparent. Can be picked up throughout space in the radio-frequency and microwave range. Was emitted about 380,000 years after the Big Bang.

Dark energy: Little-understood force that is thought to lead to a more rapid expansion of the universe. Today dark energy accounts for about 70 percent of the total energy in the universe.

Dark matter: Unknown form of matter whose existence can only be deduced from the effect of its gravity on the universe. It is thought to account for about 85 percent of the total mass of the universe.

Doppler effect: Describes the shifting of colors/frequencies of light as a result of the relative motion of two objects. In astronomy, movements along the line of sight can be measured using this effect.

EHT (Event Horizon Telescope): Global VLBI network of millimeter-wave radio telescopes that captured the first image of a black hole.

Electromagnetic waves: Radiation without mass that moves at the speed of light in a vacuum. Examples include light, infrared or thermal radiation, microwaves, and radio waves, as well as x-ray and gamma rays.

Entropy: Measure of disorder in a system. Without the addition of energy, the entropy in a system can only increase.

ERC (European Research Council): EU institution that finances basic research by excellent scientists.

ESA (European Space Agency): The EU's space agency builds space telescopes and operates satellites.

ESO (European Southern Observatory): Operates optical telescopes in Chile like the VLT and the La Silla Observatory and is also a partner in ALMA and APEX.

Event Horizon: Invisible border surrounding a black hole beyond which matter, radiation, and all information falls irrevocably into the black hole.

Exoplanet: A planet orbiting a star other than the sun.

Fourier transform: Mathematical operation that converts waves into their frequencies and vice versa. Used in radio interferometry to produce images because it measures "image frequencies."

Gaia: Spacecraft and telescope launched by the ESA to map the stars in our Milky Way.

Galactic center: Center of the Milky Way, located 26,000 light-years from Earth.

Galaxy: System of hundreds of billions of stars, planets, and gas nebulae that are gravitationally linked and revolve around a center. Our home galaxy is the Milky Way.

General theory of relativity: Theory authored by Albert Einstein that describes the relationship between space, time, and gravity. Mass warps, or curves, space, and the curved space determines the movement of mass and the passing of time.

Globular clusters: Mostly old, spherical groupings of up to 100,000 stars that are gravitationally linked and orbit galaxies.

GLT (Greenland Telescope): A 12-meter telescope in Greenland, and part of the Event Horizon Telescope (EHT) as well as the Global mm-VLBI Array.

GPS (global positioning system): Network of satellites used to determine locations on Earth.

Gravitational lens: According to general relativity, the gravitational lens effect is seen to occur whenever light is deflected by the influence of a very massive object. If light waves pass by a massive object on their way to Earth—for example, galaxies, stars, or a black hole—the waves do not pass by it in a straight line, but rather are deflected and curved. When this happens, effects can be seen that are similar to those caused by an optical lens made of glass, and that make it possible to draw conclusions about the form and mass of the gravitational lens.

Gravity: Mutual force of attraction exerted by bodies of mass. Described in general relativity by the deforming of space-time.

GRAVITY: Interferometer operated by the ESO, which connects four telescopes of the VLT and makes high-resolution near-infrared images, for example, of the stars in the Galactic Center.

GRMHD (general relativistic magnetohydrodynamics): Method for simulating the movements of gas within the magnetic fields around black holes.

Hawking radiation: Model attributed to the physicist Stephen Hawking that states that black holes can gradually evaporate as a result of quantum effects. As yet unconfirmed by experiment.

Haystack Observatory: MIT's radio observatory in Westford, Massachusetts.

Hubble-Lemaître Law: States that as a result of the universe's expansion, galaxies move more rapidly away from us the farther away from us they are. Can be used in connection with redshifting and spectroscopy to measure distances in space.

Hubble Space Telescope: Powerful spacecraft operated by NASA and the ESA that has observed outer space in several ranges of the electromagnetic spectrum, from infrared, to visible light, to the ultraviolet range.

Interferometry: Technique based on the layering of waves. In radio astronomy you can combine radio waves received by different telescopes and produce high-resolution images from the interference patterns (see *VLBI, radio interferometer*).

IRAM (Institut de Radioastronomie Millimétrique): German, French, and Spanish research association. It operates the NOEMA telescope in France (2,600 meters above sea level) and the 30-meter telescope on Pico del Veleta in Spain (2,920 meters above sea level)—both are part of the EHT.

ISS (International Space Station): The only constantly occupied space station in space; orbits 400 kilometers above the Earth.

JCMT (James Clerk Maxwell Telescope): Radio telescope in Hawaii operating in the submillimeter wave range; part of the EHT.

Jet: Concentrated, hot stream of plasma that is shot out by the magnetic fields of certain cosmic objects. The jet of a supermassive black hole shoots out at nearly the speed of light and extends as far as millions of light-years into space.

Light speed: 299,792.458 km/s. It is always constant. Neither information nor matter can be transported faster than light speed.

Light-year: Distance that light moving in a vacuum travels in one year: 1 light-year = 0.307 parsecs = 9.46047×10^{12} km.

LMT (Large Millimeter Telescope): Fifty-meter radio telescope in Mexico located at an elevation of 4,593 meters above sea level on the dormant Sierra Negra volcano; part of the EHT.

LOFAR (Low-Frequency Array): European radio interferometry network of 30,000 low-frequency radio antennas that searches for signals from the early phase of the universe. The center of operations is located in the Netherlands.

Max Planck Society: A large, elite research institution in Germany with affiliate institutes in several scientific fields.

Messier 87 (M87): Giant elliptical galaxy located 55 million light-years from Earth. The supermassive black hole at its center is the first that EHT astronomers were able to capture an image of. First cataloged by Charles Messier.

Milky Way: Our own, disk-shaped galaxy with a spiral structure. Contains between 200 and 400 billion stars. The sun orbits the center of the Milky Way once every 200 million years.

Millimeter waves: Radio waves in the frequency range between approximately 43 GHz and 300 GHz, with a wave length between 1 and 10 millimeters.

MIT (Massachusetts Institute of Technology): Renowned polytechnic university in Cambridge, Massachusetts.

NASA (National Aeronautics and Space Administration): The United States' space agency.

Neutron star: Collapsed, ultracompact star that weighs about as much as the sun but measures only about 20 to 25 kilometers in diameter and is made up of neutrons (see *atom*). The end phase of development for many massive stars.

NRAO (National Radio Astronomy Observatory): US research organization that operates (or participates in the joint operation of) various radio telescopes—among them ALMA, the VLA, and the VLBA.

NSF (National Science Foundation): US agency responsible for financing research projects.

Nuclear fusion: By melting atomic nuclei together—primarily of hydrogen atoms, which then form helium—stars produce energy.

Parallax: Apparent shifting in position of a celestial object when it is observed from two different locations. Using this effect and the astronomical unit, it's possible to measure the distance of stars from Earth.

Parsec (parallax second): Astronomical unit of length that is the equivalent of about 3.26 light-years or 206,000 astronomical units. The term traces back to the measurement of distance using the parallax of stars.

Photons: Light particles found in electromagnetic radiation. Light at all wavelengths can be both a wave and a particle.

Planet: Spherical object made of gas or rock that orbits the sun with minimal impediment. Doesn't produce its own radiation via nuclear fusion; rather, it only reflects sunlight. Our solar system has eight planets (Mercury, Venus, Earth, Mars, Jupiter, Saturn, Uranus, Neptune). Planets that orbit other stars are called exoplanets.

Plasma: Ultrahot gas consisting of protons and electrons, in which atoms are broken up into their individual components.

Protostar: Young star in the development phase.

Pulsar: Rapidly rotating neutron star that sends out radio waves like a lighthouse and flashes at regular intervals.

Quantum physics: Describes physical systems in which certain conditions can only take on particular (discrete/quantized) values. Primarily applies to the smallest elementary particles.

Quasar (quasi-stellar radio source): Active nucleus (see *black hole*) of a very distant galaxy that emits a large amount of radiation and is known for its high luminosity.

Radboud University: University in Nijmegen, in the eastern Netherlands, founded as a Catholic university in 1925.

Radio interferometer: Network of radio telescopes that synchronously observe the same celestial object in order to achieve a higher resolution, equivalent to the resolution of a telescope as large as the distance between the two most-distant telescopes in the network.

Radio Telescope Effelsberg: One-hundred-meter radio telescope in the Eifel mountains that is operated by the Max Planck Institute for Radio Astronomy in Bonn.

Red giant: Bloated, aging star; nuclear fusion only occurs in a layer around its core. The star swells and emits red light.

Redshift: As a result of the universe's expansion and of the rapid movement of galaxies away from us, light is shifted to colors with longer wavelengths, or "redder" colors (see *Doppler effect*). The light from the edge of a black hole is also redshifted because of the severe curvature of space-time.

Sagittarius A* (Sgr A*): Compact radio source at the Galactic Center, likely the supermassive black hole in the center of our Milky Way weighing 4 million solar masses and standing at a distance of 26,000 light-years from Earth.

SAO (Smithsonian Astrophysical Observatory): Astronomical research institution in Cambridge, Massachusetts.

SETI (search for extraterrestrial intelligence): Blanket term for programs that began to appear in the 1960s and are seeking to discover life in outer space.

Singularity: Place behind the event horizon of a black hole, in which the curvature of space-time is infinite and mass becomes concentrated. The earliest stage in the formation of the universe is referred to as the Big Bang singularity or the initial singularity.

SMA (Submillimeter Array): Interferometer consisting of eight radio telescopes that is part of the EHT network. It is located on the Mauna Kea volcano in Hawaii at 4,115 meters above sea level.

Solar mass: Standard astronomical unit of mass; 2×10^{30} kg.

Special theory of relativity: One of Einstein's theories of relativity, which describes variations in time and distance as a result of relative motions. Unlike the general theory of relativity, it does not take gravity into account and is important when nearing light speed.

Spectroscopy: Method for measuring light in which light is separated out into its individual colors (its spectrum). As a result of processes described by quantum physics, atoms of different elements absorb or emit light within narrow color ranges and can be identified based on these colors. Radial velocities can be measured via redshifting and the Doppler effect.

SPT (South Pole Telescope): The 10-meter radio telescope at the Amundsen-Scott South Pole Station in Antarctica is part of the EHT network and is located at an elevation of 2,817 meters above sea level.

Star: Hot ball of gas that produces energy via nuclear fusion. The sun is also a star. The larger and heavier a star is, the hotter it is and the shorter its life span.

Supernova: Very bright explosion of a massive star at the end of its life.

Synchrotron radiation: Electromagnetic radiation that is produced by the diversion of electrons moving at nearly the speed of light through a magnetic field. Describes the radio emission of black holes.

VATT (Vatican Advanced Technology Telescope): Optical telescope operated by the Vatican Observatory on Mount Graham.

Venus transit: Passage of Venus in front of the sun. By measuring the phenomenon, it became possible to calculate the distance between the Earth and the sun (the astronomical unit).

VLA (Very Large Array): Radio interferometer consisting of twenty-seven 25-meter radio telescopes in New Mexico spread out over a distance of up to 36 kilometers.

VLBA (Very Long Baseline Array): VLBI network in the US consisting of ten 25-meter antennas with a distance between them of up to 8,600 kilometers. The equivalent in Europe is the European VLBI Network, or EVN.

VLBI (Very Long Baseline Interferometry): Method of interferometric measurement in which radio telescopes located at a great distance from one another are linked up and observe a radio source at the same time. The actual image is formed later on the computer.

VLT (Very Large Telescope): Observatory with four individual 8-meter telescopes operated by the ESO on the Cerro Paranal in Chile at 2,850 meters above sea level.

White dwarf: After nuclear fusion is extinguished, most old stars end up as compact, approximately Earth-size crystal spheres weighing about one solar mass. At first they are very hot and burn a bluish-white color, but they cool down over a long period of time.

White hole: Hypothetical area of space-time that represents the opposite of a black hole and disgorges mass instead of attracting it.

Wormhole (Einstein-Rosen Bridge): Potential link between two distant areas of space-time. This "tunnel" is theoretically allowed for by the general theory of relativity, but might not exist.

Additional information and astronomical terms can be found at: **https://www.einstein-online.info** (under Useful/Dictionary)

Image Credits

Notes

Prologue: And We Really Can See Them

1   Live stream of the EU press conference in Brussels: https://youtu.be/Dr20f19czeE. ESO press release: https://www.eso.org/public/germany/news/eso1907. Video zooming in on the black hole: https://www.eso.org/public/germany/videos/eso 1907c. NSF press conference: https://www.youtube.com/watch?v=lnJi0Jy692w.

2   See photo insert, Figure 1, page 1.

Chapter 1: Humankind, the Earth, and the Moon

1   The air density in low Earth orbit is $5 \times 10^{-9} \text{g/cm}^3$, as opposed to a normal air density of 1,204 kg/m³ (10^{-3}g/cm^3): Kh. I. Khalil and S. W. Samwel, "Effect of Air Drag Force on Low Earth Orbit Satellites During Maximum and Minimum Solar Activity," *Space Research Journal* 9 (2016): 1–9, https://scialert.net/fulltext/?doi =srj.2016.1.9.

2   Ethan Siegel, "The Hubble Space Telescope Is Falling," Starts with a Bang, *Forbes*, October 18, 2017, https://www.forbes.com/sites/startswithabang/2017/10/18 /the-hubble-space-telescope-is-falling/#71ac8b1b7f04; Mike Wall, "How Will the Hubble Space Telescope Die?" Space.com, April 24, 2015, https://www.space .com/29206-how-will-hubble-space-telescope-die.html.

3   Job 26:7 (King James Version).

4   Psalms 90:4 (KJV).

5   S. M. Brewer, J.-S. Chen, A. M. Hankin, E. R. Clements, C. W. Chou, D. J. Wineland, D. B. Hume, and D. R. Leibrandt, "^{27}Al+ Quantum-Logic Clock with a Systematic Uncertainty below 10^{-18}," *Physical Review Letters* 123 (2019): 033201, https://ui .adsabs.harvard.edu/abs/2019PhRvL.123c3201B.

6   Rømer and Huygens used the orbit of the moon Io around Jupiter as a clock and determined that this clock ran a bit slow when the Earth was farther away from

Jupiter in its orbit around the sun than a few months before. The light from Jupiter arrives a few minutes later than expected; the Io clock lags behind.

7   Michelson was born in Prussia and moved to the United States with his parents at age two: https://www.nobelprize.org/prizes/physics/1907/michelson /biographical.

8   It's not certain, however, that Einstein was influenced by the Michelson-Morley experiment in a crucial way. The near relativity of electromagnetism was probably more important. See Jeroen van Dongen, "On the Role of the Michelson-Morley Experiment: Einstein in Chicago," *Archive for History of Exact Sciences* 63 (2009): 655–63, https://ui.adsabs.harvard.edu/abs/2009ar Xiv0908.1545V.

9   "Andre and Marit's Moon bounce wedding," YouTube, February 15, 2014, https:// www.youtube.com/watch?v=RH3z8TwGwrY.

10 Adam Hadhazy, "Fact or Fiction: The Days (and Nights) Are Getting Longer," *Scientific American*, June 14, 2010, https://www.scientificamerican.com/article /earth-rotation-summer-solstice.

11 M. P. van Haarlem, and 200 contributors, "LOFAR: The Low Frequency Array," *Astronomy and Astrophysics* 556 (2013): A2.

Chapter 2: The Solar System and Our Evolving Model of the Universe

1   P. K. Wang and G. L. Siscoe, "Ancient Chinese Observations of Physical Phenomena Attending Solar Eclipses," *Solar Physics* 66 (1980): 187–93, https:// doi.org/10.1007/BF00150528; also see https://eclipse.gsfc.nasa.gov/SEhistory /SEhistory.html#-2136.

2   Yuta Notsu, et al., "Do Kepler Superflare Stars Really Include Slowly Rotating Sun-like Stars?: Results Using APO 3.5 m Telescope Spectroscopic Observations and Gaia-DR2 Data," *The Astrophysical Journal* 876 (2019): 58, https://ui.adsabs .harvard.edu/abs/2019ApJ...876...58N.

3   Tweet by Mark McCaughrean, @markmccaughrean, January 5, 2020, https:// twitter.com/markmccaughrean/status/1213827446514036736.

4   Knowledge of many Stone Age artifacts (the Lascaux cave, carvings on an eagle bone in the Dordogne, Stonehenge, the lunar map at Knowth) remains vague and contested. See Karenleigh A. Overmann, "The Role of Materiality in Numerical Cognition," *Quaternary International* 405 (2016): 42–51, https://doi.org/10.1016/j .quaint.2015.05.026; P. J. Stooke, "Neolithic Lunar Maps at Knowth and Baltinglass, Ireland," *Journal for the History of Astronomy* 25, no. 1 (1994): 39–55, https://doi.org/10.1177/002182869402500103. Nevertheless, human curiosity argues against assumptions that humans only began studying the sky with the advent of verifiable written sources.

5   Jörg Römer, "Als den Menschen das Mondfieber packte," *Der Spiegel*, July 16, 2019, https://www.spiegel.de/wissenschaft/mensch/mond-in-der-achaeologie -zeitmesser-der-steinzeit-a-1274766.html.

6   The International Celestial Reference System (ICRS) is a system of coordinates that is compiled from Very Long Baseline Interferometry (VLBI) observations of quasars; the orientation of the Earth in space within this system is determined according to International Earth Rotation and Reference Systems Services (IERS) Earth Orientation Parameters. It can be used, for example, to connect coordinates on Earth in the International Terrestrial Reference System (ITRS) with satellite coordinates: https://www.iers.org/IERS/EN/Science/ICRS/ICRS.html.

7   John Steele, *A Brief Introduction to Astronomy in the Middle East* (London: Saqi, 2008). Scholars of the ancient Middle East have encountered evidence of ersatz kings. In Mesopotamia, at the time of a solar or lunar eclipse, a puppet ruler was placed on the throne in place of the king who had been marked by an evil omen. A prisoner or mentally challenged person was chosen for the purpose. During this time the real king lived as a simple peasant. Only after a hundred days had passed did the priests give the all clear.

8   Matthew 2:1–13 (KJV). Nowhere in the biblical text is it actually stated that they are kings, or that there are three of them. Usage and historical context make it probable that the figures referred to here are astrologically trained experts. More details can be found in my WordPress blog post on the subject (Heino Falcke, "The Star of Bethlehem: A Mystery (Almost) Resolved?" October 28, 2014, https:// hfalcke.wordpress.com/2014/10/28/the-star-of-bethlehem-a-mystery-almost -resolved) and in the literature cited in the post, in particular, George H. van Kooten and Peter Barthel, eds., *The Star of Bethlehem and the Magi: Interdisciplinary Perspectives from Experts on the Ancient Near East, the Greco-Roman World, and Modern Astronomy* (The Hague: Brill Academic Publishers, 2015).

9   Bede, *De Natura Rerum*; Johannes de Sacro Bosco (b. 1230 AD), *Tractatus de Sphaera*, see http://www.bl.uk/manuscripts/Viewer.aspx?ref=harley_ms_3647_f024r.

10   John Freely, *Before Galileo: The Birth of Modern Science in Medieval Europe* (New York: Overlook Press, 2014).

11   Sebastian Follmer, "Woher haben die Wochentage ihre Namen," *Online Focus*, September 11, 2018, https://praxistipps.focus.de/woher-haben-die-wochentage -ihre-namen-alle-details_96962.

12   The astronomy of the Indian astronomer Aryabhata (b. 476 AD) was geocentric but posited that the Earth rotated; see Kim Plofker, *Mathematics in India* (Princeton: Princeton University Press, 2009). For more on Indian astronomy, see N. Podbregar, "Jantar Mantar: Bauten für den Himmel," scinexx.de, September 15, 2017, https://www.scinexx.de/dossier/jantar-mantar-bauten -fuer-den-himmel.

13   Joseph Needham, with the research assistance of Wang Ling, *Science and Civilisation in China: Vol. 2, History of Scientific Thought* (Cambridge: Cambridge University Press, 1956), cited in "The Chinese Cosmos: Basic Concepts," Asia for Educators, http://afe.easia.columbia.edu/cosmos/bgov/cosmos.htm.

14   For example, see Peter Harrison, *The Territories of Science and Religion* (Chicago: University of Chicago Press, 2015). A summary written by the author can be

found here: https://theologie-naturwissenschaften.de/en/dialogue-between
-theology-and-science/editorials/conflict-myth.

15 Another such myth, also well-known from film, is the murder of Hypatia by a
Christian mob and the burning down of the Library of Alexandria. It doesn't
diminish Hypatia's significance as a courageous, wise woman to note that her
case doesn't serve to support the "science versus Christianity" thesis. The murder
was more political in nature—the library didn't really exist anymore as such—
and besides all that, there is scant factual evidence. See Charlotte Booth, *Hypatia:
Mathematician, Philosopher, Myth* (Stroud, UK: Fonthill, 2016). See also Maria
Dzielska, "Hypatia wird zum Opfer des Christentums stilisiert," *Der Spiegel*,
April 25, 2010, https://www.spiegel.de/wissenshaft/mensch/interview-zum-film
-agora-hypatia-wird-zum-opfer-des-christentums-stilisiert-a-690078.html; and
further, Cynthia Haven, "The Library of Alexandria—Destroyed by an Angry Mob
with Torches? Not Very Likely," The Book Haven (blog), March 2016, https://
bookhaven.stanford.edu/2016/03/the-library-of-alexandria-destroyed-by-an
-angry-mob-with-torches-not-very-likely.

16 Hans Lipperhey from Middelburg is generally regarded as the inventor of the
telescope, but there are others who claimed this invention as their own.

17 Mario Livio, *Galileo and the Science Deniers* (New York, Simon & Schuster, 2020).
For a contrasting view, see this review of the book: Thony Christie, "How to
Create Your Own Galileo," The Renaissance Mathmeticus (blog), May 27, 2020,
https://thonyc.wordpress.com/2020/05/27/how-to-create-your-own-galileo.
Christie shows that much of our present-day image of Galileo has been
poeticized, and pulls no punches in his criticism of Livio's book.

18 Ulinka Rublack, *Der Astronom und die Hexe: Johannes Kepler und seine Zeit*
(Stuttgart: Klett-Cotta, 2019).

19 Newton was a professor of theology and was known among his peers as an
outstanding Bible scholar, though he also explored alchemical and heretical
ideas in secret. Robert Iliffe, "Newton's Religious Life and Work," The Newton
Project, http://www.newtonproject.ox.ac.uk/view/contexts/CNTX00001.

20 In Episode IV, Han Solo proudly declares that he once made the Kessel run in 12
parsecs. This sounds like a description of time, but in the opinion of some fans is
meant to be an indication of distance. See https://jedipedia.fandom.com/wiki
/parsec. Astronomers always shift uncomfortably in their seats whenever they
hear this line.

21 Alberto Sanna, Mark J. Reid, Thomas M. Dame, Karl M. Menten, and Andreas
Brunthaler, "Mapping Spiral Structure on the Far Side of the Milky Way," *Science*
358 (2017): 227–30, https://ui.adsabs.harvard.edu/abs/2017Sci...358..227S.

Chapter 3: Einstein's Happiest Thought

1 This might be more of a philosophical question, but a completely empty space
has zero entropy and does not develop; thus, there's no time to measure either. A

completely empty space without matter or vacuum energy would be nothingness in the truest sense of the word, and physics has nothing to say about it—though mathematics might.

2   *Light* is used here in a more general sense and includes all forms of interactions that are mostly conducted at the speed of light. Space has no meaning in a hypothetical universe with matter that never interacts. Here the question presents itself as to what we should call reality. The solutions to Einstein's field equations exist just as well without there having to be any light or matter in space-time. Of course, then space and time are reduced to a purely mathematical concept described by the term *nothingness*.

3   For example, Philip Ball, "Why the Many-Worlds Interpretation Has Many Problems," *Quanta Magazine*, October 18, 2018, https://www.quantamagazine .org/why-the-many-worlds-interpretation-of-quantum-mechanics-has-many -problems-20181018; Robbert Dijkgraaf, "There Are No Laws of Physics. There's Only the Landscape," *Quanta Magazine*, June 4, 2018, https://www.quanta magazine.org/there-are-no-laws-of-physics-theres-only-the-landscape -20180604.

4   The process by which quantum states experience loss of information on their way to becoming macroscopic objects is generally described by the concept of decoherence. A more thorough, broadly accessible treatment of quantum physics can be found, for example, in Claus Kiefer, *Der Quantenkosmos: Von der zeitlosen Welt zum expandierenden Universum* (Frankfurt: S. Fischer, 2008).

5   There are historical reasons for the fact that physicists speak in terms of the speed of light. From a modern point of view it would also be possible to name this absolute maximum speed the "speed of gravity," after gravitational waves, or better yet the "speed of causality." In the theory of relativity, the speed of light is a fundamental quality of space-time, namely the natural relationship of the spatial scale to the timescale.

6   J. C. Hafele and Richard E. Keating, "Around-the-World Atomic Clocks: Predicted Relativistic Time Gains," *Science* 177 (1972): 166–68, https://ui.adsabs.harvard .edu/abs/1972Sci...177..166H. What's important is that all three clocks are moving—relative to a nonrotational "inertial system," like the center of the Earth or the fixed stars! At the equator, a clock on the ground moves east at about 1,600 km/h. If we fly east in an Airbus A330 traveling 900 km/h, then our velocity is the velocity of the plane plus the Earth's rotation, up to 2,500 km/h. Flying west, we move 900 km/h slower relative to the center of the Earth than the Earth's surface—so just about 700 km/h, but ultimately still traveling east! The Mr. Clock that flew east was traveling fastest relative to the center of the Earth and thus time passed most slowly for it, relatively speaking. The Mr. Clock that flew west moved slowest, relatively speaking, and so time passed most quickly. The clock that waited dutifully on the ground also wasn't standing still relative to the center of the Earth. It provides us with a reference time and ticked slower than a clock at the center would have, faster than the clock flying east, and slower than

the clock flying west. Thus, the experiment indeed tests aspects of the general theory of relativity and the equivalence principle.

7   R. Malhotra, Matthew Holman, and Thomas Ito, "Chaos and Stability of the Solar System," *Proceedings of the National Academy of Science* 98, no. 22 (2001): 12342–43, https://ui.adsabs.harvard.edu/abs/2001PNAS...9812342M.

8   My colleague Paul Groot was our department head for many years.

9   The physicist and mathematician Pierre-Simon Laplace was responsible for an important step forward in the development of celestial mechanics with his 1823 work *Traité de mécanique céleste*. The mathematician Urbain Le Verrier succeeded in predicting Neptune's existence by studying disruptions in Uranus's orbit in 1846.

10 Einstein started as a third-tier employee, but he'd been promoted by the time he published the theory.

11 Hanoch Gutfreund and Jürgen Renn, *The Road to Relativity: The History and Meaning of Einstein's "The Foundation of General Relativity"* (Princeton: Princeton University Press, 2015).

12 Pauline Gagnon, "The Forgotten Life of Einstein's First Wife," *Scientific American*, December 19, 2016, https://blogs.scientificamerican.com/guest-blog/the -forgotten-life-of-einsteins-first-wife. A somewhat different portrayal is offered in Allen Esterson and David C. Cassidy, contribution by Ruth Lewin Sime, *Einstein's Wife: The Real Story of Mileva Einstein-Maric* (Boston: MIT Press, 2019).

13 From personal letters, as quoted in Gutfreund, H. und J. Renn, *The Road to Relativity: The History and Meaning of Einstein's "The Foundation of General Relativity"* (Princeton: Princeton University Press, 2015), 57.

14 Albert Einstein, "How I Created the Theory of Relativity," reprinted in: Y. A. Ono, *Physics Today* 35, no. 8 (1982): 45, https://physicstoday.scitation.org/doi/10 .1063/1.2915203.

15 Strictly speaking, the equivalence principle only applies to a point mass, since in this example Einstein's feet would be pulled down to Earth with a tiny bit more force than his head. This is a result of what are called tidal forces. Earth is comparatively small, so the effect is minimal. While falling into a small black hole, however, Einstein would definitely notice something; in fact he'd be spaghetti-fied.

16 A nice test of the equivalence principle has been performed with radio-astronomical measurements of a pulsar in a triple-star system with two white dwarfs: https://www.mpg.de/14921807/allgemeine-relativitaetstheorie-pulsar; G. Voisin, et al., "An Improved Test of the Strong Equivalence Principle with the Pulsar in a Triple-Star System," *Astronomy & Astrophysics* 638 (2020): A24, https://www.aanda.org/articles/aa/abs/2020/06/aa38104-20/aa38104 -20.html.

17 Hanoch Gutfreund and Jurgen Renn, *The Road to Relativity: The History and Meaning of Einstein's "The Foundation of General Relativity"* (Princeton: Princeton University Press, 2015).

18 Daniel Kennefick, "Testing Relativity from the 1919 Eclipse: A Question of Bias," *Physics Today* 62, no 3. (2009): 37, https://physicstoday.scitation.org/doi/10.1063/1.3099578.

19 The light is deflected half by the curvature of space and half by the curvature of time. The latter is already accounted for in Newton's theory, which therefore predicts half the value of the deflection.

20 J.-F. Pascual-Sánchez, "Introducing Relativity in Global Navigation Satellite Systems," *Annalen der Physik* 16 (2007): 258–73, https://ui.adsabs.harvard.edu/abs/2007AnP...519..258P. By a simple calculation, an error of 39 microseconds per day equals a positioning error of about 10 kilometers. This is stated in many popular articles, but it's not clear whether this applies to the actual system, where all the satellite clocks are making a comparable error. More precise calculations are in the works (M. Pössel and T. Müller, in progress).

21 A good overview of general relativistic effects as they relate to GPS can be found in this article: Neil Ashby, "Relativity in the Global Positioning System," *Living Reviews in Relativity* 6 (2003): article no. 1, https://link.springer.com/article/10.12942/lrr-2003-1.

22 With thanks to Jun Ye for this tip. E. Oelker, et al., "Optical Clock Intercomparison with $6 \times 10-1^9$ Precision in One Hour," arXiv eprints (February 2019), https://ui.adsabs.harvard.edu/abs/2019arXiv190202741O.

Chapter 4: The Milky Way and Its Stars

1   See *Spectroscopy* in the glossary.

2   Joshua Sokol, "Stellar Disks Reveal How Planets Get Made," *Quanta Magazine*, May 21, 2018, https://www.quantamagazine.org/stellar-disks-reveal-how-planets-get-made-20180521.

3   A small number of the hydrogen atoms in us were probably never in stars, but rather have been drifting through space in diffuse gas since the Big Bang.

4   Originally, the planet Dimidium was called "51 Pegasi b." This is also the name that most astronomers would recognize.

5   J. E. Enriquez, et al., "The Breakthrough Listen Initiative and the Future of the Search for Intelligent Life," *American Astronomical Society Meeting Abstracts* 229 (2017): 116.04, https://ui.adsabs.harvard.edu/abs/2017AAS...22911604E.

Chapter 5: Dead Stars and Black Holes

1   G. W. Collins, W. P. Claspy, and J. C. Martin, "A Reinterpretation of Historical References to the Supernova of AD 1054," *Publications of the Astronomical Society of the Pacific* 111, no. 761 (1999): 871–80, https://ui.adsabs.harvard.edu/abs/1999PASP..111..871C.

2   Some researchers do link the Chaco Canyon pictograph with the supernova of 1054, which appeared on July 4, 1054, in the eastern part of the constellation

Taurus: https://www2.hao.ucar.edu/Education/SolarAstronomy/supernova
-pictograph. Doubt has recently been cast on this interpretation, however: Clara
Moskowitz, "'Supernova' Cave Art Myth Debunked," *Scientific American*, January
16, 2014, https://blogs.scientificamerican.com/observations/e28098supernovae
28099-cave-art-myth-debunked.

3   Ingrid H. Stairs, "Testing General Relativity with Pulsar Timing," *Living Reviews in
Relativity* 6 (2003): 5, https//ui.adsabs.harvard.edu/abs/2003LRR.....6....5S.

4   M. Kramer and I. H. Stairs, "The Double Pulsar," *Annual Review of Astronomy
and Astrophysics* 46 (2008): 541–72, https://ui.adsabs.harvard.edu/abs/2008
ARA&A..46..541K.

5   Andreas Brunthaler found the supernova SN 2008iz by chance in his data.

6   N. Kimani, et al., "Radio Evolution of Supernova SN 2008iz in M 82," *Astronomy
and Astrophysics* 593 (2016): A18, https://ui.adsabs.harvard.edu/abs/2016
A&A...593A..18K.

7   J. R. Oppenheimer and G. M. Volkoff, "On Massive Neutron Cores," *Physical
Review* 55, no. 374 (1939): 374—but neutron stars were first proposed by Baade
and Zwicky: W. Baade and F. Zwicky, "Remarks on Super-Novae and Cosmic
Rays," *Physical Review* 46 (1934): 76–77, https://ui.adsabs.harvard.edu/abs
/1934PhRv...46...76B.

8   Schwarzschild probably didn't find the solution in Russia, but rather on the
western front in the southern Vosges, as a letter to Arnold Sommerfeld makes
clear: https://leibnizsozietaet.de/wp-content/uploads/2017/02/Kant.pdf.

9   A few months later the Dutch scientist Johannes Droste independently found an
even more elegant solution—which was roundly ignored, because Droste had
only published it in Dutch. At this time it was still important to be able to
communicate in German.

10   Hanoch Gutfreund and Jurgen Renn, *The Road to Relativity: The History and
Meaning of Einstein's "The Foundation of General Relativity"* (Princeton: Princeton
University Press, 2015).

11   "LEXIKON DER ASTRONOMIE: Schwarzschild-Lösung," https://spektrum.de
/lexikon/astronomie/schwarzschild-loesung/431.

12   As I was writing I thought I'd come up with something really original in using a
river analogy to describe a black hole, but apparently somebody's already written
a whole academic article about it: Andrew J. S. Hamilton and Jason P. Lisle, "The
River Model of Black Holes," *American Journal of Physics* 76 (2008): 519–32,
https://ui.adsabs.harvard.edu/abs/2008AmJPH..76..519H.

An entire collection of visual models depicting the general theory of relativity
can be found here: Markus Pössel, "Relatively Complicated? Using Models to
Teach General Relativity at Different Levels," arXiv eprints (December 2018):
1812.11589, https://ui.adsabs.harvard.edu/abs/2018arXiv181211589P.

13   Jeremy Bernstein, "Albert Einstein und die Schwarzen Löcher," *Spektrum der
Wissenschaft*, August 1, 1996, https://www.spektrum.de/magazin/albert-einstein
-und-die-schwarze-loecher/823187.

14 Here a point does not mean a point in space in the sense found in the general theory of relativity. The central singularity is a boundary of infinitely curved space-time.

15 Ann Ewing, "'Black Holes' in Space," *The Science News-Letter* 85, no. 3 (January 18, 1964): 39, https://jstor.org/stable/3947428?seq=1.

16 Roy P. Kerr, "Gravitational Field of a Spinning Mass as an Example of Algebraically Special Metrics," *Physical Review Letters* 11 (1963): 237–38, https://ui.adsabs .harvard.edu/abs/1963PhRvL..11..237K.

17 This effect is a significant factor in the formation of plasma jets around black holes, though not absolutely necessary. It is known under the term *Blandford-Znajek process* and is a variant of the Penrose process, in which rotational energy can be extracted from the black hole with the help of light or particles.

18 Information about the Africa millimeter-wave telescope here: https://www.ru .nl/astrophysics/black-hole/africa-millimetre-telescope; M. Backes , et al., "The Africa Millimetre Telescope," *Proceedings of the 4th Annual Conference on High Energy Astrophysics in Southern Africa* (HEASA 2016): 29, https://ui.adsabs .harvard.edu/abs/2016heas.confE..29B.

19 "Mistkäfer orientieren sich an der Milchstraße," Spiegel Online, January 24, 2013, https://www.spiegel.de/wissenschaft/natur/mistkaefer-orientieren-sich-an-der -milchstrasse-a-879525.html.

20 Dirk Lorenzen, "Die Beobachtung der Andromeda-Galaxie," Deutschlandfunk, October 5, 2018, https://www.deutschlandfunk.de/vor-95-jahren-die -beobachtung-der-andromeda-galaxie.732.de.html?dram:article_id=429694.

21 Trimble, V., "The 1920 Shapley-Curtis Discussion: Background, Issues, and Aftermath." *Publications of the Astronomical Society of the Pacific* 107, no. 718 (1995): 1133, https://ui.adsabs.harvard.edu/abs/1995PASP.107.1133T.

22 E. P. Hubble, *The Realm of the Nebulae* (New Haven: Yale University Press, 1936). Available online at: https://ui.adsabs.harvard.edu/abs/1936rene.book.....H.

23 M. J. Reid and A. Brunthaler, "The Proper Motion of Sagittarius A*. III. The Case for a Supermassive Black Hole," *The Astrophysical Journal* 892 (2020): 39, https:// ui.adsabs.harvard.edu/abs/2020ApJ...616..872R.

Chapter 6: Galaxies, Quasars, and the Big Bang

1 Emilio Elizalde, "Reasons in Favor of a Hubble-Lemaître-Slipher's (HLS) Law," *Symmetry* 11 (2019): 15, https://ui.adsabs.harvard.edu/abs/2019Symm...11...35E.

2 Richard Porcas was the last person to photograph the 90-meter telescope in Green Bank. The photo has long hung in the hallway of the MPIfR in Bonn.

3 Ken Kellermann, "The Road to Quasars" (lecture, Caltech Symposium: "50 Years of Quasars," September 9, 2013), https://sites.astro.caltech.edu/q50/pdfs/Kellermann.pdf.

4 Maarten Schmidt, "The Discovery of Quasars" (lecture, Caltech Symposium: "50 Years of Quasars," September 9, 2013), https://sites.astro.caltech.edu/q50 /Program.html.

Chapter 7: The Galactic Center

1   Charles H. Townes and Reinhard Genzel, "Das Zentrum der Galaxis," *Spektrum der Wissenschaft*, June 1990, https://www.spektrum.de/magazin/das-zentrum -der-galaxis/944605.

2   Pronounced "Sadge A Star."

3   When too much material falls toward a black hole, so much radiation is produced that the gas is blown away by the radiation pressure. The maximum limit of mass accretion is called the Eddington limit.

4   Heino Falcke and Peter L. Biermann, "The Jet-Disk Symbiosis. I. Radio to X-ray Emission Models for Quasars," *Astronomy and Astrophysics* 293 (1995): 665–82, https://ui.adsabs.harvard.edu/abs/1995A&A...293..665F.

5   Heino Falcke and Peter L. Biermann, "The Jet/Disk Symbiosis. III. What the Radio Cores in GRS 1915+105, NGC 4258, M 81, and SGR A* Tell Us About Accreting Black Holes," *Astronomy and Astrophysics* 342 (1999): 49–56, https://ui.adsabs .harvard.edu/abs/1999A&A...342...49F.

6   Roland Gredel, ed., *The Galactic Center, 4th ESO/CTIO Workshop*, ASPC 102 (1996), http://www.aspbooks.org/a/volumes/table_of_contents/?book_id=214.

7   A. Eckart and R. Genzel, "Observations of Stellar Proper Motions Near the Galactic Centre," *Nature* 383 (1996): 415–17, https://ui.adsabs.harvard.edu/abs /1996Natur.383..415E.

8   B. L. Klein, A. M. Ghez, M. Morris, and E. E. Becklin, "2.2μm Keck Images of the Galaxy's Central Stellar Cluster at 0.05 Resolution," *The Galactic Center* 102 (1996): 228, https://ui.adsabs.harvard.edu/abs/1996ASPC..102..228K.

9   A. M. Ghez, M. Morris, E. E. Becklin, A. Tanner, and T. Kremenek, "The Accelerations of Stars Orbiting the Milky Way's Central Black Hole," *Nature* 407 (2000): 349–51, https://ui.adsabs.harvard.edu/abs/2000Natur.407..349G.

10 Karl M. Schwarzschild, Mark J. Reid, Andreas Eckart, and Reinhard Genzel, "The Position of Sagittarius A*: Accurate Alignment of the Radio and Infrared Reference Frames at the Galactic Center," *The Astrophysical Journal* 475 (1997): L111–14, https://ui.adsabs.harvard.edu/abs/1997Ap...475L.111M.

11 M. J. Reid and A. Brunthaler, "The Proper Motion of Sagittarius A*. II. The Mass of Sagittarius A*," *The Astrophysical Journal* 616 (2004): 872–84, https://ui.adsabs .harvard.edu/abs/2004ApJ...616..872R.

12 R. Schödel, et al., "A Star in a 15.2-Year Orbit Around the Supermassive Black Hole at the Centre of the Milky Way," *Nature* 419 (2002): 694–96, https://ui.adsabs .harvard.edu/abs/2002Natur.419..694S.

13 L. Meyer, et al., "The Shortest-Known-Period Star Orbiting Our Galaxy's Supermassive Black Hole," *Science* 338 (2012): 84, https://ui.adsabs.harvard.edu /abs/2012Sci...338...84M.

14 R. Genzel, et al., "Near-Infrared Flares from Accreting Gas Around the Supermassive Black Hole at the Galactic Centre," *Nature* 425 (2003): 934–37, https://ui.adsabs.harvard.edu/abs/2003Natur.425..934G.

15 F. K. Baganoff, et al., "Rapid X-Ray Flaring from the Direction of the Supermassive Black Hole at the Galactic Centre," *Nature* 413 (2001): 45–48, https://ui.adsabs .harvard.edu/abs/2001Natur.413...45B.

16 Gravity Collaboration and R. Abuter, et al., "Detection of Orbital Motions Near the Last Stable Circular Orbit of the Massive Black Hole Sgr A*," *Astronomy and Astrophysics* 618 (2018): L10, https://ui.adsabs.harvard.edu/abs/2018 A&A...618L..10G.

17 Geoffrey C. Bower, Melvyn C. H. Wright, Heino Falcke, and Donald C. Backer, "Interferometric Detection of Linear Polarization from Sagittarius A* at 230 GHz," *The Astrophysical Journal* 588 (2003): 331–37, https://ui.adsabs.harvard .edu/abs/2003ApJ...588..331B.

18 H. Falcke, E. Körding, and S. Markoff, "A Scheme to Unify Low-Power Accreting Black Holes: Jet-Dominated Accretion Flows and the Radio/X-Ray Correlation," *Astronomy and Astrophysics* 414 (2004): 895–903, https://ui.adsabs.harvard.edu /abs/2004A&A...414..895F.

19 F. Yuan, S. Markoff, and H. Falcke, "A Jet-ADAF Model for Sgr A*," *Astronomy and Astrophysics* 383 (2002): 854–63, https://ui.adsabs.harvard.edu/abs/2002 A&A...383..854Y.

Chapter 8: The Idea Behind the Image

1 John 20:29 (KJV).

2 The image resolution of a telescope is expressed in angular units, here in radians (rad): 2π rad equals 360°. The formula expresses, in terms of their angle to the line of sight, how far apart two points of light have to be in order to be distinguishable.

3 Alan E. E. Rogers, et al., "Small-Scale Structure and Position of Sagittarius A* from VLBI at 3 Millimeter Wavelength," *Astrophysical Journal Letters* 434 (1994): L59, https://ui.adsabs.harvard.edu/abs/1994ApJ...434L.59R.

4 T. P. Krichbaum, et al., "VLBI Observations of the Galactic Center Source SGR A* at 86 GHz and 215 GHz," *Astronomy and Astrophysics* 335 (1998): L106–10, https://ui.adsabs.harvard.edu/abs/1998A&A...335L.106K.

5 Heino Falcke, et al., "The Simultaneous Spectrum of Sagittarius A* from 20 Centimeters to 1 Millimeter and the Nature of the Millimeter Excess," *The Astrophysical Journal* 499 (1998): 731–34, https://ui.adsabs.harvard.edu/abs /1998ApJ...499..731F.

6 H. Falcke, et al., "The Central Parsecs of the Galaxy: Galactic Center Workshop" (proceedings of a meeting held in Tucson, Arizona, September 7–11, 1998), https://ui.adsabs.harvard.edu/abs/1999ASPC..186.....F.

7 J. A. Zensus and H. Falcke, "Can VLBI Constrain the Size and Structure of SGR A*?," *The Central Parsecs of the Galaxy*, ASP Conference Series 186 (1999): 118, https://ui.adsabs.harvard.edu/abs/1999ASPC..186..118Z.

8 A nice visualization of the paths the light takes can be found here: T. Müller

and M. Pössel, "Ray tracing eines Schwarzen Lochs und dessen Schatten," Haus der Astronomie, http://www.haus-der-astronomie.de/3906466 /BlackHoleShadow.

9   Tilman Sauer and Ulrich Majer, eds., *David Hilbert's Lectures on the Foundations of Physics 1915–1927* (Springer Verlag, 2009). See also: M. von Laue, *Die Relativitätstheorie* (Friedrich Vieweg & Sohn, 1921), 226.

10 C. T. Cunningham and J. M. Bardeen, "The Optical Appearance of a Star Orbiting an Extreme Kerr Black Hole," *The Astrophysical Journal* 183 (1973): 237–64, https://ui.adsabs.harvard.edu/abs/1973ApJ...183..237C; J.-P. Luminet, "Image of a Spherical Black Hole with Thin Accretion Disk," *Astronomy and Astrophysics* 75 (1979): 228–35, https://ui.adsabs.harvard.edu/abs/1979A&A....75..228L; S. U. Viergutz, "Image Generation in Kerr Geometry. I. Analytical Investigations on the Stationary Emitter-Observer Problem," *Astronomy and Astrophysics* 272 (1993), https://ui.adsabs.harvard.edu/abs/1993A&A...272..355V. For the first article the calculations and drawings were done by hand, for the second the calculations were done on the computer and the drawings done by hand, and for the third both calculations and drawings were done on the computer.

11 Later, Professor Ferdinand Schmidt-Kaler, who at that time was trying to support my work and to whom I'm grateful for recommending me for the Akademiepreis of the Berlin-Brandenburg Academy of Sciences, informed me that a former student of his, just a few weeks after us and completely independent of us, had also entered the term *shadow* of a black hole into the literature—albeit in a very abstract and mathematical paper. A. de Vries, "The Apparent Shape of a Rotating Charged Black Hole, Closed Photon Orbits, and the Bifurcation Set A_4," *Classical and Quantum Gravity* 17 (2000): 123–44, https://ui.adsabs.harvard.edu/abs /2000CQGra..17..123D.

12 Heino Falcke, Fulvio Melia, and Eric Agol, "Viewing the Shadow of the Black Hole at the Galactic Center," *The Astrophysical Journal* 528 (2000): L13–16, https://ui.adsabs.harvard.edu/abs/2000ApJ...528L..13F.

13 Heino Falcke, Fulvio Melia, and Eric Agol, "The Shadow of the Black Hole at the Galactic Center," *American Institute of Physics Conference Series* 522 (2000): 317–20, https://ui.adsabs.harvard.edu/abs/2000AIPC..522..317F.

14 Press release, "First Image of a Black Hole's 'Shadow' May Be Possible Soon," Max Planck Institute for Radio Astronomy in Bonn, January 17, 2000, http:// www3.mpifr-bonn.mpg.de/staff/junkes/pr/pr1_en.html.

Chapter 9: Building a Global Telescope

1   The Max Planck Institute in Bonn and the Steward Observatory had built the Heinrich-Hertz Telescope (HHT), a 10-meter dish, on Mount Graham in Arizona together. When the Germans pulled out a few years later, it was renamed the Submillimeter Telescope (SMT) and the University of Arizona was working with great initiative to try to keep it alive on their own. In Hawaii there was the James

Clerk Maxwell Telescope (JCMT) on Mauna Kea, a 15-meter dish. Today astronomers from China, Korea, Japan, and the Academia Sinica in Taipei, among others, work alongside each other at the JCMT. The two European telescopes operated by the Institut de Radioastronomie Millimétrique (IRAM) on Pico del Veleta in Spain and the Plateau de Bure in the French Alps were on firm footing and continued to be operated on a permanent basis. Other observatories were only in the planning phase, among them the Large Millimeter Telescope (LMT) in Mexico—ideally located for us, geographically speaking. It was supposed to be a 50-meter supertelescope, but the time when it was brought online was delayed until 2011, and even then it wasn't completely finished. Even on the south pole a telescope specially built for cosmology was in the planning, which became operational in 2007. But it was another eight years before my colleague Dan Marrone from Arizona and his colleagues were able to successfully link up the telescope in the remoteness of Antarctica with a VLBI network.

2   H. Falcke, et al., "Active Galactic Nuclei in Nearby Galaxies," *American Astronomical Society Meeting Abstracts* 200 (2002): 51.06, https://ui.adsabs .harvard.edu/abs/2002AAS...200.5106F.

3   P. A. Shaver, "Prospects with ALMA," in: R. Bender and A. Renzini, eds., *The Mass of Galaxies at Low and High Redshift: Proceedings of the European Southern Observatory and Universitäts-Sternwarte München Workshop Held in Venice, Italy, 24–26 October 2001* (Springer-Verlag, 2003), 357, https://ui.adsabs.harvard.edu /abs/2003mglh.conf.357S.

4   *De Gelderlander*, April 2003.

5   G. C. Bower, et al., "Detection of the Intrinsic Size of Sagittarius A* Through Closure Amplitude Imaging," *Science* 304 (2004): 704–8, https://ui.adsabs.harvard .edu/abs/2004Sci...304..704B.

6   S. Markoff, et al., eds., "GCNEWS–Galactic Center Newsletter," vol. 18, http:// www.aoc.nrao.edu/~gcnews/gcnews/Vol.18/editorial.shtml.

7   The minutes are in my private archive. My Chilean colleague Neil Nagar also took part occasionally.

8   Sheperd S. Doeleman, et al., "Event-Horizon-Scale Structure in the Supermassive Black Hole Candidate at the Galactic Centre," *Nature* 455 (2008): 78–80, https:// ui.adsabs.harvard.edu/abs/2008Natur.455...78D.

9   *A Science Vision for European Astronomy* (Garching: ASTRONET, 2010), 27.

10 Sheperd Doeleman, et al., "Imaging an Event Horizon: submm-VLBI of a Super Massive Black Hole," *Astro2010: The Astronomy and Astrophysics Decadal Survey* 68 (2009), https://ui.adsabs.harvard.edu/abs/2009astro2010S.68D.

11 Monika Mościbrodzka, et al., "Radiative Models of SGR A* from GRMHD Simulations," *The Astrophysical Journal* 706 (2009): 497–507, https://ui.adsabs .harvard.edu/abs/2009ApJ...706..497M.

12 Monika Mościbrodzka, Heino Falcke, Hotaka Shiokawa, and Charles F. Gammie, "Observational Appearance of Inefficient Accretion Flows and Jets in 3D GRMHD

Simulations: Application to Sagittarius A*," *Astronomy and Astrophysics* 570 (2014): A7, https://ui.adsabs.harvard.edu/abs/2014A&A...570A...7M.

13 Monika Mościbrodzka, Heino Falcke, and Hotaka Shiokawa, "General Relativistic Magnetohydrodynamical Simulations of the Jet in M 87," *Astronomy and Astrophysics* 586 (2016): A38, https://ui.adsabs.harvard.edu/abs/2016A&A...586A ..38M. But Dexter's work, too, had already provided an excellent prediction based on GRMHD simulations: Jason Dexter, Jonathan C. McKinney, and Eric Agol, "The Size of the Jet Launching Region in M87," *Monthly Notices of the Royal Astronomical Society* 421 (2012): 1517–28, https://ui.adsabs.harvard.edu/abs /2012MNRAS.421.1517D.

14 In the end, because the chances were so bad, 50 percent fewer applications were submitted in our round, so the chances were actually 3 percent.

15 Images and videos from our ERC project can be found at: https://blackholecam .org. C. Goddi, et al., "BlackHoleCam: Fundamental Physics of the Galactic Center," *International Journal of Modern Physics D* 26 (2017): 1730001–239, https://ui.adsabs.harvard.edu/abs/2017IJMPD..2630001G.

16 R. P. Eatough, et al., "A Strong Magnetic Field Around the Supermassive Black Hole at the Centre of the Galaxy," *Nature* 501 (2013): 391–94, https://ui.adsabs .harvard.edu/abs/2013Natur.501..391E.

17 Doctoral students: Michael Janßen (Lower Rhine), Sara Issaoun (Canada), Freek Roelofs, Jordy Davelaar, Thomas Bronzwaer, Christiaan Brinkerink (Netherlands), Raquel Fraga-Encinas (Spain), Shan Shan (China); postdoc: Cornelia Müller (Germany); senior scientists: Ciriaco Goddi (Italy), Monika Mos´cibrodzka (Poland), Daan van Rossum (Germany); project manager: Remo Tilanus (Netherlands).

18 L. D. Matthews, et al., "The ALMA Phasing System: A Beamforming Capability for Ultra-High-Resolution Science at (Sub)Millimeter Wavelengths," *Publications of the Astronomical Society of the Pacific* 130 (2018): 015002, https://ui.adsabs .harvard.edu/abs/2018PASP..130a5002M.

19 The idea for the melody probably came from chief operator Bob Moulton, but Tom Folkers, who wrote the operating system for the entire SMT, was the one who programmed it.

20 Tweets and images from February 11, 2016, when after a thesis defense we watched the press conference for the LIGO/Virgo Collaboration in the auditorium at Radboud University: https://twitter.com/hfalcke/status /697819758562041857?s=21; https://twitter.com/hfalcke/status/697805820 143276033?s=21.

21 Interview with Karsten Danzmann on Deutschlandfunk, February 12, 2016, https://www.deutschlandfunk.de/gravitationswellen-nachweis-einstein-hatte -recht.676.de.html?dram:article_id=345433.

22 Mickey Steijaert, "The Rising Star of Sara Issaoun," *Vox: Independent Radboud University Magazine*, June 21, 2019, https://www.voxweb.nl/international /the-rising-star-of-sara-issaoun.

Chapter 10: Striking Out on Expedition

1   See photo insert and glossary for the telescopes in the EHT experiment: ALMA and APEX in the Atacama Desert in Chile, SMT on Mount Graham in Arizona, the James Clerk Maxwell Telescope and the Submillimeter Array on Mauna Kea in Hawaii, the IRAM 30-meter Telescope on Pico del Veleta, the Large Millimeter Telescope (LMT) on the dormant volcano of Sierra Negra in Mexico, and the South Pole Telescope (SPT) at the Amundsen-Scott South Pole Station. The SPT cannot observe the M87 Galaxy because it is located in the northern part of the sky.

2   Pink Floyd, "Comfortably Numb," track 6 on *The Wall*, Harvest Records, 1979. Lyrics from: https://de.wikipedia.org/wiki/Roger_Waters.

3   This time Michael Janßen goes to Mexico with computer scientist Katie Bouman from MIT. My Italian colleague Ciriaco Goddi travels with Geoff Crew from Haystack to ALMA in Chile. Remo Tilanus flies to Hawaii to work alongside Mareki Honma and other colleagues from Asia at the JCMT. Sara Issaoun again looks after the telescope in Arizona, along with Freek Roelofs and Junhan Kim, who had prepared the South Pole Telescope over Christmas.

4   Peter Mezger was the director of the submillimeter wave group at the Max Planck Institute for Radioastronomy. His book *Blick in das kalte Weltall* was published in 1992 with the story of the telescopes, particularly the SMT/HHT.

5   Thomas Krichbaum from Bonn and Rebecca Azulay, a young Spanish postdoc working at the MPI, along with the two Spaniards Pablo Torne and Salvador Sánchez from IRAM. Torne specializes in astronomical observation and Sánchez in the technical equipment. The station director Carsten Kramer was also there with us at the beginning.

6   Actually two movies were made: *The Edge of All We Know* by Peter Galison of Harvard, www.blackholefilm.com, and *How to See a Black Hole: The Universe's Greatest Mystery* by Henry Fraser, Windfall Films, both initiated by the Harvard Group.

7   M. J. Valtonen, et al., "A Massive Binary Black-Hole System in OJ 287 and a Test of General Relativity," *Nature* 452 (2008): 851–53, https://ui.adsabs.harvard.edu/abs/2008Natur..452.851V.

8   Andrew Nadolski was the second man at the south pole.

9   Karl Schuster, director, IRAM.

10 David Hughes, director of the LMT.

11 Lizzie Wade, "Violence and Insecurity Threaten Mexican Telescopes," *Science*, February 6, 2019, https://www.sciencemag.org/news/2019/02/violence-and-insecurity-threaten-mexican-telescopes#.

Chapter 11: An Image Resolves

1   The calibration team includes Lindy Blackburn and Maciek Wielgus from Harvard University, Chi-kwan Chan from Arizona, my doctoral students Sara Issaoun and Michael Janßen, and Ilse van Bemmel from Dwingeloo.

2   A. R. Thompson, J. M. Moran, and G. W. Swenson, *Interferometry and Synthesis in Radio Astronomy, 3rd Edition,* (Springer Verlag, 2017).

3   Radboud Pipeline for the Calibration of High Angular Resolution Data: M. Janßen, et al., "rPICARD: A CASA-Based Calibration Pipeline for VLBI Data. Calibration and Imaging of 7mm VLBA Observations of the AGN Jet in M 87," *Astronomy and Astrophysics* 626 (2019): A75, https://ui.adsabs.harvard.edu /abs/2019A&A...626A..75J. The JIVE team under Mark Kettenis and Ilse van Bemmel also took part, along with Kazi Rygl and Elisabetta Liuzzo from Bologna.

4   A young team led by Michael Johnson, Katie Bouman, and Kazunori Akiyama heads up the imaging group. Harvard PhD student Andrew Chael also takes part. On the European side, Thomas Krichbaum and José Luis Gómez from Spain also play a strong role. In all, more than fifty scientists are involved. Sara Issaoun is among them, and even the theorist Monika Mościbrodzka tries her hand at imaging.

5   Bouman and Johnson's group at Harvard make up one team. I was in Team II with my PhD students Freek Roelofs, Michael Janßen, and Sara Issaoun. Thomas Krichbaum and José Luis Gómez from Spain and their colleagues made up the third team, which specialized on the CLEAN algorithm. A young group of our Asian colleagues under Keiichi Asada made up the fourth team.

6   FITS: Flexible Image Transport System.

7   H. Falcke, "How to Make the Invisible Visible" (lecture, TEDxRWTH Aachen, 2018), https://www.youtube.com/watch?v=ZHeBi4e9xoM.

8   Images from the EHT imaging workshop at Harvard in 2017 can be found here: https://eventhorizontelescope.org/galleries/eht-imaging-workshop-october -2017.

9   These are two "regularized maximum likelihood" (RML) methods (eht-imaging and SMILI) and the CLEAN algorithm.

10   Chi-kwan Chan led a group to determine the color scale.

11   Francis Reddy, "NASA Visualization Shows a Black Hole's Warped World," nasa .gov, September 25, 2019, https://www.nasa.gov/feature/goddard/2019/nasa -visualization-shows-a-black-hole-s-warped-world.

12   EHT theory groups formed around Charles Gammie in Illinois, Ramesh Narayan at Harvard, Luciano Rezzolla in Frankfurt, and Monika Mościbrodzka in Nijmegen.

13   Under the leadership of Feryal Özel in Arizona, Keiichi Asada in Japan, Jason Dexter in Garching, and Avery Broderick at the Perimeter Institute in Canada. Meanwhile Christian Fromm from the BlackHoleCam team in Frankfurt developed a new "genetic algorithm" in order to estimate parameters of the black hole by comparing the images with simulations.

14   Photos and videos from the collaboration meeting in November 2018 in Nijmegen: https://www.ru.nl/astrophysics/black-hole/event-horizon-telescope -collaboration-0/eht-collaboration-meeting-2018.

15   The EHT publication committee was led by Laurent Loinard from Mexico and

my Dutch colleague Huib Jan van Langevelde, as well as Ramesh Narayan and John Wardle from the US.

16 Yosuke Mizuno, et al., "The Current Ability to Test Theories of Gravity with Black Hole Shadows," *Nature Astronomy* 2 (2018): 585–90, https://ui.adsabs.harvard .edu/abs/2018NatAs...2..585M.

17 "UH Hilo Professor Names Black Hole Capturing World's Attention," press release, University of Hawai'i, April 10, 2019, https://www.hawaii.edu/news/2019 /04/10/uh-hilo-professor-names-black-hole.

18 Video zooming in on the black hole: https://www.eso.org/public/germany /videos/eso1907c.

19 Nik's music video, with cellphone videos taken at the press conference and the black hole image: [Nik], "Wahrscheinlich" (music video), https:/www.youtube .com/watch?v=oaUBCDpsFCw.

20 The alert astroblogger was Daniel Fischer, https://skyweek.lima-city.de—thanks! Also Ralf Nestler from *Der Tagesspiegel* notified us.

Chapter 12: Beyond the Powers of Our Imagination

1   L. L. Christensen, et al., "An Unprecedented Global Communications Campaign for the Event Horizon Telescope First Black Hole Image," *Communicating Astronomy with the Public Journal* 26 (2019): 11, https://ui.adsabs.harvard.edu /abs/2019CAPJ...26...11C.

2   Google Doodle: https://www.google.com/doodles/first-image-of-a-black-hole.

3   Tim Elfrink, "Trolls Hijacked a Scientist's Image to Attack Katie Bouman. They Picked the Wrong Astrophysicist," *The Washington Post*, April 12, 2019, https:// www.washingtonpost.com/nation/2019/04/12/trolls-hijacked-scientists-image -attack-katie-bouman-they-picked-wrong-astrophysicist.

4   L. L. Christensen, et al., "An Unprecedented Global Communications Campaign."

5   Th. Rivinius, "A Naked-Eye Triple System with a Nonaccreting Black Hole in the Inner Binary," *Astronomy and Astrophysics* 637 (2020): L3, https://ui.adsabs .harvard.edu/abs/2020A&A...637L...3R.

6   The diameter of the black hole is approximately 24 kilometers.

7   The art history of the image of the black hole is the subject of a doctoral thesis by Emilie Skulberg at Cambridge.

8   M. Backes, et al., "The Africa Millimetre Telescope," *Proceedings of the 4th Annual Conference on High Energy Astrophysics in Southern Africa (HEASA 2016)*: 29, https://ui.adsabs.harvard.edu/abs/2016heas.confE..29B.

9   Freek Roelofs, et al., "Simulations of Imaging the Event Horizon of Sagittarius A* from Space," *Astronomy and Astrophysics* 625 (2019): A124, https://ui.adsabs .harvard.edu/abs/2019A&A . . . 625A.124R; Daniel C. M. Palumbo, et al., "Metrics and Motivations for Earth-Space VLBI: Time-Resolving Sgr A* with the Event Horizon Telescope," *The Astrophysical Journal* 881 (2019): 62, https://ui.adsabs .harvard.edu/abs/2019ApJ...881...62P.

Chapter 13: Beyond Einstein?

1   Event Horizon Telescope Collaboration, et al., "First M87 Event Horizon Telescope Results. I. The Shadow of the Supermassive Black Hole," *Astrophysical Journal Letters* 875 (2019): L1, https://ui.adsabs.harvard.edu/abs/2019 ApJ...875L...1E.

2   The hypothesis that antimatter falls just the same as matter is currently being experimentally tested at CERN: Michael Irving, "Does Antimatter Fall Upwards? New CERN Gravity Experiments Aim to Get to the Bottom of the Matter," *New Atlas*, November 5, 2018, https://newatlas.com/cern-antimatter-gravity -experiments/57090.

3   Dennis Overbye, "How to Peer Through a Wormhole," *New York Times*, November 13, 2019, https://www.nytimes.com/2019/11/13/science/wormholes-physics -astronomy-cosmos.html.

4   For examples of information-based theories of gravity, see: Martijn Van Calmthout, "Tug of War Around Gravity," Phys.org, August 12, 2019, https:// phys.org/news/2019-08-war-gravity.html; Stephen Wolfram, "Finally We May Have a Path to the Fundamental Theory of Physics . . . and It's Beautiful," stephenwolfram.com (blog), https://writings.stephenwolfram.com/2020/04 /finally-we-may-have-a-path-to-the-fundamental-theory-of-physics-and-its -beautiful; Tom Campbell, et al., "On Testing the Simulation Theory," *International Journal of Quantum Foundations* 3 (2017): 78–99, https://www .ijqf.org/archives/4105; M. Keulemans, "Leven we eigenlijk in een hologram? Het zou zomaar kunnen," *de Volkskrant*, March 10, 2017, https://www.volkskran .nl/wetenschap/leven-we-eigenlijk-in-een-hologram-het-zou-zomaar-kunnen ~bb4boda3/.

5   Actually if you stirred for an infinitely long time in a big bowl of alphabet soup, you could randomly produce a book, but there would be no way of telling that you had, and it would disappear again immediately—you'd have to stop at exactly the right moment. It's more efficient to write a book than to wait for one to suddenly appear.

6   Ethan Siegel, "Ask Ethan: What Was the Entropy of the Universe at the Big Bang?" *Forbes*, April 15, 2017, https://www.forbes.com/sites/startswithabang/2017/04 /15/ask-ethan-what-was-the-entropy-of-the-universe-at-the-big-bang.

7   In quantum physics one describes the preservation of information in a quantum system, i.e., the development of its wave function, with the term *unitarity* and the process of measuring a quantum particle often as the *collapse of the wave function*. The "condition" of a quantum particle and/or its wave function only determines the probability with which a certain value is measured. Before every measurement of quantum particles, you can only precisely measure the most probable value—i.e., the mean value over several measurements. But once a value is measured, it remains constant until something else is measured. Multiple measurements thus change the values of particles.

8   "Schwarze Löcher erinnern sich an ihre Opfer," Spiegel Online, March 9, 2004,
    https://www.speigel.de/wissenschaft/weltall/hawking-verliert-wette-schwarze
    -loecher-erinnern-sich-an-ihre-opfer-a-289599.html.

9   Even in isolated quantum systems without gravity, it could be that information
    is thermalized and becomes lost, if the calculations described in this article are
    correct: Maximilian Kiefer-Emmanouilidis, et al., "Evidence for Unbounded
    Growth of the Number Entropy in Many-Body Localized Phases," *Physical
    Review Letters* 124 (2020): 243601, https://journals.aps.org/prl/abstract/10
    .1103/PhysRevLett.124.243601.

Chapter 14: Omniscience and Limitations

1   Jeremiah 33:22 (KJV).

2   John Horgan, *The End of Science* (New York: Little, Brown, 1997).

3   Ethan Siegel, "No Galaxy Will Ever Truly Disappear, Even in a Universe with Dark
    Energy," *Forbes*, March 4, 2020, https://www.forbes.com/sites/startswithabang
    /2020/03/04/no-galaxy-will-ever-truly-disappear-even-in-a-universe-with-dark
    -energy.

4   Sam Harris, *Free Will* (New York: Free Press, 2012), 5 (Kindle version): "Free will *is*
    an illusion. Our wills are simply not of our own making. Thoughts and intentions
    emerge from background causes of which we are unaware and over which we
    exert no conscious control. We do not have the freedom we think we have.
    Free will is actually more than an illusion (or less), in that it cannot be made
    conceptually coherent. Either our wills are determined by prior causes and we
    are not responsible for them, or they are the product of chance and we are not
    responsible for them."

5   In this context scientists have also begun debating the concept of emergence.

6   An example for readers versed in mathematics: I determine the frequency of a
    light wave in a flat space with the help of a Fourier transform. But this only
    provides an infinitely exact value if I integrate the wave from $-\infty$ to $+\infty$; then, for
    example, the Fourier transform of a sine function is the exact same as the delta
    function. If I have less time than eternity, then the frequency of even a perfect
    sine function is always imprecise. For the same reason I can only measure the
    point in time or the location of an event with infinite precision if I have an
    infinite amount of frequencies or wavelengths at my disposal. But because every
    event and every particle is always spatially and chronologically finite, it is also in
    fact always imprecise.

7   Natalie Wolchover, "Does Time Really Flow?: New Clues Come from a Century-
    Old Approach to Math," *Quanta Magazine*, April 7, 2020, https://www.quanta
    magazine.org/does-time-really-flow-new-clues-come-from-a-century-old
    -approach-to-math-20200407.

8   Lawrence Krauss, *A Universe from Nothing: Why There Is Something Rather than
    Nothing* (New York: Atria Books, 2014): Pos. 104/3284 (Kindle version).

9   For this reason the entropy at the beginning of the universe was actually lower than it is now, when energy and mass are widely distributed throughout space. Every individual star, planet, or person might seem more "orderly" than the Big Bang, but seen in the context of the entire universe that scarcely makes a difference. It's like with the box full of toy blocks in the playroom: at the moment of the Big Bang everything was in a small box; now everything's in a giant playroom. Even if you take a few of the blocks and build a nice little house here and there, the big picture is one of immense disorder.

10 With the probable exception of "dark energy," which could be an energy of empty space.

11 Martin Rees, *Just Six Numbers: The Deep Forces That Shape the Universe* (New York: Basic Books, 2001).

12 K. Landsman, "The Fine-Tuning Argument," arXiv eprints (May 2015): 1505.05359, https://ui.adsabs.harvard.edu/abs/2015arXiv150505359L.

13 I would have liked to discuss the subject with him, but at least we can still read his thoughts on it: Stephen Hawking, *Brief Answers to the Big Questions* (London: John Murray, 2018).

14 John 1:1. The verse reads, in full: "In the beginning was the Word, and the Word was with God, and the Word was God." (KJV).

15 Genesis 11:1–9, the building of the Tower of Babel. In this famous story God first has to descend in order to be able to look at the tower.

16 1 Corinthians 13:13 (Easy-to-Read Version), Paul's song in praise of love.

Index